Coexistentialism and the Unbearable Intimacy of Ecological Emergency

Ecocritical Theory and Practice

Series Editor: Douglas A. Vakoch, California Institute of Integral Studies, USA

Ecocritical Theory and Practice highlights innovative scholarship at the interface of literary/cultural studies and the environment, seeking to foster an ongoing dialogue between academics and environmental activists.

Titles in the Series

Coexistentialism and the Unbearable Intimacy of Ecological Emergency, by Sam Mickey
Disability and the Environment in American Literature: Toward an Ecosomatic Paradigm, edited by Matthew J.C. Cella
The Mythology of the Animal Farm in Children's Literature: Over the Fence, by Stacy E. Hoult-Saros
Ecocriticism in Taiwan: Identity, Environment, and the Arts, edited by Chia-ju Chang and Scott Slovic
Ecotheology in the Humanities: An Interdisciplinary Approach to Understanding the Divine and Nature, edited by Melissa Brotton
T.S. Eliot, Poetry, and Earth: The Name of the Lotos Rose, by Etienne Terblanche

Coexistentialism and the Unbearable Intimacy of Ecological Emergency

Sam Mickey

LEXINGTON BOOKS
Lanham • Boulder • New York • London

Published by Lexington Books
An imprint of The Rowman & Littlefield Publishing Group, Inc.
4501 Forbes Boulevard, Suite 200, Lanham, Maryland 20706
www.rowman.com

Unit A, Whitacre Mews, 26-34 Stannary Street, London SE11 4AB

British Library Cataloguing in Publication Information Available

Library of Congress Cataloging-in-Publication Data

Names: Mickey, Sam, 1981- author.
Title: Coexistentialism and the unbearable intimacy of ecological emergency / Sam Mickey.
Description: Lanham : Lexington Books, 2016. | Series: Ecocritical theory and practice | Includes bibliographical references and index.
Identifiers: LCCN 2016021958 (print) | LCCN 2016023679 (ebook) | ISBN 9781498517652 (cloth : alk. paper) | ISBN 9781498517676 (Electronic)
Subjects: LCSH: Existentialism. | Human ecology.
Classification: LCC B819 .M535 2016 (print) | LCC B819 (ebook) | DDC 142/.78--dc23
LC record available at https://lccn.loc.gov/2016021958

∞™ The paper used in this publication meets the minimum requirements of American National Standard for Information Sciences Permanence of Paper for Printed Library Materials, ANSI/NISO Z39.48-1992.

Printed in the United States of America

Table of Contents

Acknowledgments

In a book about coexistence, it is impossible to give thorough acknowledgments of all those who supported the inception and development of this project. Who or what could be left out? Acknowledging everything and everyone is impossible, and the "Acknowledgments" section of a book is not necessarily the place to attempt to let the impossible happen, so I will restrict myself to a few acknowledgments for people who most intimately supported this project. I have to begin with very special thanks for my parents, Cindy and Bob Mickey. Without their coexistence, I would not exist. Additionally, I must thank my father for introducing me to existentialism at a young age, teaching me how to laugh at paradox and the absurd, and I cannot express enough thanks to my sister Sabrina for laughing with me so much. It's absurd. Another close ally in my encounters with existentialism is a courageously uncanny artist, Patrick Cash, a friend who I carry in my memory.

I am grateful for the support of the faculty, staff, and students of the various institutions at which I teach as an adjunct professor in and around San Francisco, including the University of San Francisco, the California Institute of Integral Studies, and Dominican University of California. As an interdisciplinary scholar, I am happy to have found support at those institutions for my work across boundaries of philosophy, religion, and environmental studies. Thanks to Sean Kelly, Adam Robbert, Justin Hayes, Whitney Bauman, Clayton Crockett, Catherine Keller, Brian Swimme, Aaron Weiss, Elizabeth McAnally, Jeremy Watts, Jesse Mills, Matt Segall, George James, and Keith Wayne Brown. Your conversations are still inspiring and informing me. I am indebted to Mary Evelyn Tucker, John Grim, and everyone involved with the Forum on Religion and Ecology at Yale for providing a very hospitable context for my work, and for opening new passageways in ecological humanities. Gratitude is also due to the Pacific Association for the Continental Tradition, which is at the growing edge of continental philosophy today. Members of PACT have been gracious interlocutors during the development of the ideas in this book. In particular, special thanks go to Gerard Kuperus, Marjolein Oele, Brian Treanor, Tim Freeman, Elizabeth Sikes, Jason Wirth, Chris Lauer, and Dorothea Olkowski. I was able to participate in two workshops on the topic of place as part of the 2015 Davies Forum at the University of San Francisco, where I received stimulating feedback on this project. I owe thanks to Gerard Kuperus for organizing those workshops and to the other participants, including David Kim, Amanda Parris, some colleagues from PACT (Brian, Jason, Marjolein, and Elizabeth), as well as Irene Klaver, whose philosophical style and perspicacity deeply informs my way into coexistentialism.

My partner Kim Carfore has been a constant source of support, intellectual and emotional companionship, and much-needed reality checks throughout the development of this project. Finally, I have to thank my feline female companion, a black cat called "misses" or "Mrs." She has been close by during much of the writing of this book. She is a companion who is reshaping my heart and mind, teaching me much about the sheer strangeness of coexistence. This is for her and for every other, for all those who I do not know how to acknowledge.

ONE

Introduction: Renewing Existentialism

Once extracted from the ground, materials like petroleum, coal, and minerals are not replaced or readily renewed, at least not on a time scale that is directly relevant to humans. They are nonrenewable resources, bearing in mind that "resources" might not be the most appropriate appellation, or rather, its appropriateness might occlude a dimension of resources that cannot be appropriated or turned into property. In contrast to nonrenewable resources, renewable resources can be used indefinitely. Consider resources like wind, sunlight, flood control, and ideas. If a windmill absorbs energy from the wind, more wind is on the way, becoming renewed as the heat from the sun is continually generating differences in atmospheric pressure around the surface of Earth. If sunlight is used by photosynthesizing plants or solar panels, more sunlight is on the way, for the next five billion years or so, until the sun begins dying. If you rely on flood control provided by a forest, you are not subtracting from the flood control that the forest provides for others. If you learn an idea, you do not thereby deplete the idea or limit the access that others have to the idea.

Renewability can be thought in terms of a scale from zero renewability (e.g., oil or coal) to relatively immediate renewability (e.g., ideas, wind, or sunlight). In between are resources that can be renewed on a time scale relevant to human life, but not instantly, with some resources more quickly replaced than others. Water drawn from a river is rapidly replaced as more water flows in, whereas a mature tree takes a considerably longer time to be replaced, as another one has to grow and mature. Whether a resource renews quickly or slowly, its renewal is dependent on certain conditions. Water loses its renewability the more it is overconsumed or polluted. When water is less available, trees and crops that depend upon water also lose their renewability. When trees are not renewed, flood control can no longer be renewed. Even wind and sunlight depend upon conditions, namely the sun's ongoing activity of nuclear fusion.

Existentialism is a renewable resource. Like any idea, theory, or literary movement, if one person learns it, it is still available for others to learn it as well. However, the renewability of existentialism is far from guaranteed. Like all resources, existentialism is vulnerable to shifting conditions. One might think that the conditions required for the renewal of existentialism are books, universities, literacy, or organizations like the Society for Phenomenology and Existential Philosophy. That seems reasonable, especially if existentialism is thought of according to its typical definition as a literary and philosophical movement that took place approximately from the middle of the nineteenth century to the middle of the twentieth century. That definition is not incorrect. However, one characteristic of the movement—if "movement" is not too emphatic a word—is that it tends to put conventional definitions in question. The thinkers and writers who became known as existentialists expressed a way of life, not just a school of thought. It is a way of life that does not conform to a set norm but attempts, rather, to stay in the fray, affirming the messy conditions of actual existence in all of their uncertainty, ambiguity, unpredictability, and absurdity.

The conditions required for the renewal of existentialism are not to be found primarily in books, universities, or academic organizations but in upheavals and crises that render a stable life of conformity undesirable or even impossible. The social and environmental emergencies currently piling up around the planet are fecund compost heaps supporting a renewal of existentialism. It is a renewal that is oriented toward ecology and the ineluctable tangle of humans and nonhumans, in short, a renewal that can be called coexistentialism. Such a renewal is the topic of the present book, which is not to say that this book presents a history or objective account of coexistentialism. It is more like a training manual or a list of practical suggestions for renewing existentialism in the midst of ecological emergency. This book can be understood in terms of "renewing existentialism" in two senses. First, existentialism is renewing, recuperative, restorative, as it rejuvenates one's engagements in the irremediable struggle of intimate coexistence, the struggle to bear the unbearable weight of an increasingly interconnected world, the struggle to bear responsibility for the damaged and degraded habitats and inhabitants composing planetary coexistence. Second, the renewing power of existentialism can be realized by renewing existentialism, updating it to become an ecological existentialism—coexistentialism.

If you recommend existentialism to someone, perhaps a companion or neighbor, it could make sense to use the term "existentialism," although such an easily manageable label runs the risk of distracting from the challenging matters addressed in the recommendation. It could also make sense to refer to some of the most famous books of existentialism, like Jean Paul Sartre's *Being and Nothingness* or Walter Kaufmann's standard anthology, *Existentialism: From Dostoevsky to Sartre*.[1] Accordingly, in what follows, I use terms like "existentialism" and "coexistentialism," and I refer to existentialist and coexistentialist writings. However, if you are not facing an emergency of ecological coexistence, those terms and books are, at best, useless, like giving car keys to someone who has no desire or need to drive. At worst, they are harmful, like giving pharmacy keys to a drug addict, granting access to professional grade jargon to help you escape your troubled existence. Books and academic labels are part of existentialism, but they are not sufficient conditions for its renewal. If everyone is satisfied with the lives they are living, then there can be no renewal of existentialism.

If you are content with the way things are, it is better to not bother learning about existentialism. This is a point that the journalist Hunter S. Thompson makes in a 1958 letter to a friend. While recommending some books on existentialism to help orient someone undergoing a challenging period in their personal life, Thompson adds the condition that a satisfied person should not engage existentialism. It is not for people who, content with their lives, have developed a curiosity about books and academic terminology. It is for those who, immersed in discontent, undertake the challenge to be.

> I'm going to steer clear of the word "existentialism," but you might keep it in mind as a key of sorts. You might also try something called *Being and Nothingness* by Jean-Paul Sartre, and another little thing called *Existentialism: From Dostoyevsky to Sartre*. These are merely suggestions. If you're genuinely satisfied with what you are and what you're doing, then give those books a wide berth. (Let sleeping dogs lie.) But back to the answer. As I said, to put our faith in tangible goals would seem to be, at best, unwise. So we do not strive to be firemen, we do not strive to be bankers, nor policemen, nor doctors. WE STRIVE TO BE OURSELVES.[2]

The present volume—*Coexistentialism*—can be thought of as an updated and expanded version of Thompson's letter. While Thompson's letter is relatively short, *Coexistentialism* will be longer, as the challenge of striving to be ourselves is becoming increasingly complicated, woven into planetary systems and globalized networks in the midst of ecological emergency, plus there are many more books to suggest and more key terms to share now than there were when Thompson wrote his letter. Furthermore, my letter has a different address than Thompson's. Whereas Thompson is addressing a friend, the reader of *Coexistentialism* is most likely a stranger or perhaps a neighbor but not necessarily a friend, or at least not yet. Thompson's friend was in the middle of some kind of personal emergency, while I am writing about the birth of coexistentialism out of the spirit of ecological emergency, specifically the current ecological emergency that is shared by all humans and, indeed, everything and everyone involved in the messy mesh of terrestrial existence.

If you do not find yourself immersed in emergency, if you are satisfied with your existence, you should follow Thompson's advice and give this book a wide berth; let the sleeping dogs of *Coexistentialism* lie, like the dreaming dingoes that figure in the "ecological existentialism" of Deborah Bird Rose.[3] How could you be satisfied with life? During a time when dingo populations are declining and, indeed, a time when a mass extinction of species is under way, the likes of which have not taken place since the extinction of the dinosaurs approximately 65 million years ago, to be satisfied with life is to utterly fail to feel one's interconnectedness in the uncertainty, uncanniness, and suffering massively distributed throughout the web of life. If you are satisfied or think you need to be satisfied, you avoid taking responsibility for your intimate intertwining with ecological emergency. If you want to face your dissatisfaction and take to heart your vulnerable engagement with emergency, then coexisentialism in its various moods, attitudes, and practices could be useful and beneficial. It would not be useful or beneficial in a conventional sense though. It would be useful and beneficial for thinking, feeling, and acting when the measures that distinguish the useful from the useless and the beneficial from the harmful are no longer working. If it is empowering, it is empowering for giving up attachments to power and control. It is empowering for giving up power and becoming possible amid the awesome turbulence of emergency. As an idea, a feeling, or a movement, coexistentialism resonates in the lyrics of what is arguably the best eco-anthem—Björk's "Jóga":

> How beautiful to be!
> State of emergency
> Is where I want to be.[4]

Where do you want to be? Do you want to affirm the emergency of our ecological situation, to say yes to an intimacy wrought with death and extinction, an intimacy too intense and uncertain to bear? Do you want to participate in the swarm of terrestrial coexistence? If so, then this term "coexistentialism" might be a useful key for you.

"We need a term like 'coexistentialism,'" as Timothy Morton puts it, "to describe what it feels like to be a swarming colony: we contain multitudes."[5] Indeed, we need a term like "coexistentialism" to provide a key that empowers participation in the swarm, like a map or compass with which one can navigate the multitudes we contain. To put it another way, coexistentialism is a fashion statement, a kind of transvestitism or drag show. It can help you come up with exuberant outfits that set the moods for your engagements in the swarm, and there are countless outfit possibilities depending on how you combine and alter different pieces. With a passion to fashion a sense of care for coexistence, this book is an attempt to get some new outfits out of the closet. The remainder of this chapter presents an overview of the renewal of existentialism as an ecological existentialism—coexistentialism—including a discussion of the main proponents of coexistentialism (including ecological humanities scholars Deborah Bird Rose and Timothy Morton) and some of the salient features that distinguish coexistentialism from existentialism. This introduction concludes with a summary of the subsequent chapters.

PASSION FOR PARADOX

Existentialism is a movement composed of the kind of people who want to think for themselves and not follow a movement. In this sense, existentialism is not just an academic or artistic movement that lasted for around a century. Existentialism can be found anywhere there are celebrations of the uniqueness and freedom of human beings. It describes a way of being that resists institutions or authorities that denigrate the singularity of existing individuals. Existentialism can be found in every refusal to live one's life according to a system. This is a point made by two of the most prominent thinkers associated with existentialism, Friedrich Nietzsche and Søren Kierkegaard. For both of them, systems are corrupt ways of being. Nietzsche puts this quite simply, with characteristic aplomb. "I distrust all systematizers and stay out of their way. The will to a system is a lack of integrity."[6] Kierkegaard strikes a similar tone. "In relation to their systems most systematizers are like a man who has built a vast palace while he himself lives nearby in a barn; they themselves do not live in the vast systematic edifice."[7]

Existentialism is about refusing to confine existence to any system or any "ism," even "existentialism." As an "ism" that is not an "ism," existentialism is characteristically paradoxical. It is a form of rebellion against ordinary opinion (*doxa*), principles of reason, and any calculations and formulas with which people turn away from reality and render life easily manageable and digestible. This is not the same as a Romantic rebellion against alienating cultural norms in favor of a harmonious relationship with nature. Existentialists seek a more radical gesture, without culture and also without nature. Even nature must be resisted. This is expressed by one of the groundbreaking contributors to existentialism, Fyodor Dostoevsky, who conveys a sense of existential rebellion in the figure of the "underground man," a self-described sick man banging his head against the wall put up by the very laws of nature. He imagines how most people would respond to him.

> "For pity's sake," they'll shout at you, "you can't rebel: it's two times two is four! Nature doesn't ask your permission; it doesn't care about your wishes, or whether you like its laws or not. You're obliged to accept it as it is, and consequently all its results as well. And so a wall is

> indeed a wall . . . etc., etc." My God, but what do I care about the laws of nature and arithmetic if for some reason these laws and two times two is four are not to my liking? To be sure, I won't break through such a wall with my forehead if I really have not got strength enough to do it, but neither will I be reconciled with it simply because I have a stone wall here and have not got strength enough.[8]

The underground man elaborates on the formula of two times two equals four. It is indicative of a formulaic life, that is, a life lived in abstraction, an existence that gives up its will to "tables and arithmetic."[9]

Even if one is not strong enough to overcome them, nature and reason must be resisted. The underground man considers the possibility that his desire for a real life free of abstraction, reason, and laws of nature is not just his own personal propensity. Perhaps it is the goal for the entire human species, a goal that makes room for the existentialist counter-formula, two times two equals five.

> And who knows (one cannot vouch for it), perhaps the whole goal mankind strives for on earth consists just in this ceaselessness of the process of achievement alone, that is to say, in life itself, and not essentially in the goal, which, of course, is bound to be nothing other than two times two is four—that is, a formula; and two times two is four is no longer life, gentlemen, but the beginning of death. . . . I agree that two times two is four is an excellent thing; but if we're going to start praising everything, then two times two is five is sometimes also a most charming little thing.[10]

The rejection of mathematics and nature is obviously absurd. It does not make sense. Existentialism is self-avowedly irrational.[11] Mathematics presents things that can be learned (*ta mathemata*), but thinkable and learnable things are not the whole story. They become lifeless immediately upon presentation, if not sooner. This is one of the factors contributing to the stereotypical experience of boredom in school. One could attribute the problem to some deficit of attention on the part of the students, but it must not be overlooked that teachers and institutional spaces holding the other pole of their attention do not exactly exhibit the kind of passion that could garner sufficient tension to hold anybody's attention. Where does such passion lie? As Kierkegaard observes, it is the nonrational, paradoxical limit of thinking that provides that passion.

> But one must not think ill of the paradox, for the paradox is the passion of thought, and the thinker without the paradox is like the lover without passion: a mediocre fellow. . . . This, then, is the ultimate paradox of thought: to want to discover something that thought itself cannot think.[12]

Dwelling in paradox, thought does not find its passion in spaces of what can be learned. It finds its passion by wavering and wandering about (*ambigere*), moving outside of that which is certain and knowable, not just outside the coordinates of reason but, more than that, outside of any perspective that would pretend to discern the coordinates of reason or to unambiguously mark the boundary between reason and nonreason. The paradoxical thinking of existentialism finds its ethics in ambiguity.[13] It might seem that ambiguity would be a grossly inadequate guide for the actions and passions of one's life. Indeed, it is a grossly inadequate guide, and that is a good thing. Relying on an adequate guide is one of the most unethical things you could do, as it amounts to disengaging from your responsibilities to others, abdicating responsibility to a determinate authority or system of rules, duties, or calculations that negotiate the ambiguity for you in advance. If you want to live your life and not have a program live your life for you, you have to wander about in the ethics of ambiguity.

An existence that faces ambiguities and paradoxes is more real, more authentic, than an existence predicated on comforting clarity and predictability. In other words, those who affirm the uncertainty of whether they are authentic are more authentic than those who think they know whether they are authentic or inauthentic. "Being a person means never being sure that you're one."[14] If you are sure that you really are a person, then there is a good chance that you are not one. You would be living in a delusion of certitude and rational control. If you do not know whether you are a person, you could be, perhaps. The question is: Where do you want to be? Do you want to be in a state of narcosis, deluded and passionless, anesthetized by formulas and certainties, or would you rather be in a state of emergency, vulnerable to the extreme urgency of passion, exposed to the unpredictable exigencies of the world? Existentialists choose the latter. A choice for the former is really no choice at all, as it involves outsourcing your choice and letting formulas, conventions, and calculations do the work for you. Living such a life is not very lively. If that is how you want to live, why not just kill yourself? This is the question, a gravely serious question, a characteristically existentialist question, asked by Albert Camus. "There is but one truly serious philosophical problem, and that is suicide. Judging whether life is or is not worth living amounts to answering the fundamental question of philosophy."[15]

Maybe there are more fundamental questions of philosophy, such as the quest for the meaning of being, the idea of goodness, or the limits of knowledge and language. However fundamental those or other questions might be, the ridiculous irrelevance of posing a question to someone who committed suicide indicates the singular importance of the problem that Camus describes. For any philosophical questions to matter for you, a choice must be made, affirming your own life, rejecting suicide. That might sound easy or obvious if one takes affirmation to be a way of ignoring the challenges and crises of existence and putting on a happy face, like an employee providing service with a smile, a naïvely optimistic person smiling in hopes for a better tomorrow, a nostalgic Romantic smiling in remembrance of an idealized past, or a spiritual-but-not-religious person wearing the detached smile of being present in the Now. I am not arguing that you cannot smile while affirming your existence. You can smile, but you do not have to. It is neither a necessary nor sufficient condition for any sort of happiness worth attaining. Forced smiles and willful ignorance might be sources of happiness, but from the perspective of someone like Camus, it is a paltry happiness compared to what is required to make a choice to reject suicide and affirm life.

Happiness in the sense of affirming one's own life means not turning away from the crisis of irrationality and meaninglessness looming over oneself and the world. Happiness is thus a close relative of the absurd. "Happiness and the absurd," as Camus puts it, are both born "of the same earth. They are inseparable."[16] Camus finds this kind of happiness exemplified in the "futile and hopeless labor" represented in the ancient Greek myth of Sisyphus.[17] For his impudent and disobedient and behavior in relation to the gods, Sisyphus—a mere mortal—is punished, condemned to the unending task of rolling a large boulder up a mountain, watching the boulder roll back down, and then going back down the mountain to repeat the process. This repeated process accomplishes nothing. It is for nothing. It is worthless, useless, and hopeless. Is it also unhappy? When Sisyphus is walking back down, on the way to bearing his burden yet again, is he depressed, angry, or at least moderately peeved? Is he resentful of the gods for the eternal repetition of the same absurd punishment? Is he regretful of the decisions he made that led to this cursed outcome? It is difficult to speculate on Sisyphus' psychological state. Perhaps it changes. He certainly has the time to go through a variety of dispositions and moods. In the reading that Camus gives, and the point here is not to debate the merits of Camus' reading for classical studies but to consider what his reading says of existentialist passions for paradox and absurdity, the answer to each of those questions is no. Camus imagines a happy Sisyphus.

For Camus, the absurd struggle of repeatedly rolling a boulder up a hill is enough to make Sisyphus happy, with no need for a god, a reason, or any such master to secure happiness.

> This universe henceforth without a master seems to him neither sterile nor futile. Each atom of that stone, each mineral flake of that night-filled mountain, in itself forms a world. The struggle itself toward the heights is enough to fill a man's heart. One must imagine Sisyphus happy.[18]

How strangely wonderful it is to be! How excellent it is to struggle for existence, to participate passionately in a world, to touch stone! If feeling a mountain beneath your feet does not make you happy, what more are you seeking? If you are seeking some reason, law, or deity to grant you happiness, there is a chance that it is precisely such seeking that is depriving you of happiness, preventing you from opening up your actual existence to atoms, minerals, stones, mountains, and countless others. This is not to say that there is no God or gods, although Camus was a self-avowed atheist. For the mythic Sisyphus, there are deities, but Sisyphus does not outsource his personal work to the divine denizens of Mount Olympus. He does not subcontract his happiness work but takes it up for himself, filling his heart through his actual struggle to be, making happiness through his absurd struggle to coexist with that stone on that night-filled mountain.

With the stone and mountain of Sisyphus or the revolt against nature by the underground man, one can detect ecologically relevant themes throughout existentialism. One could even argue that an ecological vector runs through the whole movement, moving through some thinkers more than others. For instance, on one hand, Nietzsche and Heidegger have been influential for many environmental philosophers, including deep ecologists and eco-phenomenologists, and there are numerous books that focus explicitly on the ecological implications of their thinking. That is not to say that people only have positive things to say about them. Along with appreciative ecological exegeses there are also criticisms.[19] On the other hand, other existentialists have not figured as prominently, with comparatively few works that attempt ecological readings of Kierkegaard, Dostoevsky, Sartre, or Camus. That is not to say that the latter thinkers have nothing to contribute to ecological philosophy. The happiness of Sisyphus is relevant to anyone who has a relationship to stone and mountain. The underground man's revolt against the laws of nature could be interpreted according to an "ecology without nature," such as Morton articulates.[20] Similarly, the existentialist novels and short stories of Franz Kafka can be read in terms of an ecology of things—vibrant, agentive, material beings—such as Jane Bennett articulates in her analysis of his Odradek figure—"a spool of thread who/that can run and laugh," and even speak, thus crossing boundaries between subject and object, human and inhuman, animate and inanimate.[21]

There is not a clean break between existentialism and coexistentialism. Existentialism was never totally devoid of ecological sensibilities. Its emphasis on the freedom of individual humans was never completely caged in individualism or anthropocentrism, although coexistentialists are definitely bursting the bubbles of individualism and anthropocentrism far more than existentialists. Another factor contributing to the lack of a clean break between existentialism and coexistentialism is that existentialism never stopped. Although the popular use of the term might have subsided in the middle of the twentieth century, philosophy, literature, and popular culture continued to draw from the well of existentialism. In the latter half of the twentieth century, so-called poststructuralist philosophers like Foucault, Deleuze, and Derrida combined, criticized, and complicated the predominant figures of existentialist philosophy, giving many explicit readings of Kierkegaard, Nietzsche, Heidegger, and Sartre, but also drawing on those thinkers implicitly.

The ongoing influence of existentialism is evident in popular figures like Hunter S. Thompson (mentioned above) or the filmmaker Woody Allen, whose most recent film bears an explicitly existentialist title, *Irrational Man*, and contains numerous existentialist themes in its portrayal of accidents and incidents in the life of a troubled philosophy professor, Abe (Joaquin Phoenix).[22] Existentialism can be observed in the "existentialist detectives" Bernard (Dustin Hoffman) and Vivienne (Lily Tomlin) of *I Heart Huckabees*.[23] Reverberations of existentialism echo in the comedic television series *Curb Your Enthusiasm*, in which Larry David (as himself) frequently engages in absurd situations that play with mundane cultural norms. In episode 8 ("Beloved Aunt") of the first season, David paraphrases the last sentence of *The Unnamable*, a novel by Samuel Beckett, an Irish writer and director often labeled as an existentialist (bearing in mind that the vast majority of existentialists disavow the label). David, sitting with his family in mourning after a beloved aunt died, responds to someone that says "We must go on," and his response to that sincere statement is to paraphrase, but not directly refer to, Beckett: "Must go on. Can't go on. Must go on."[24] His utterance receives no response. That scene is not just another indication of the ongoing influence of existentialism. It is also an apt summary of the history of existentialism. It is insistent and persistent, compelled to go on; at the same time, it cannot go on, as it was never a unified system or a coherent movement, and most of those associated with it did not subscribe to the label; yet, it goes on nonetheless, even when people criticize the term or the people associated with the term, even when it receives no response from friends and loved ones. The crises of late modernity have kept conditions ripe for renewing existentialism. Must go on.

COEXISTENCE

What is coexistentialism? That seems like an appropriate question to address in the opening chapter of a book that takes that noun as its title. Yet, of course, things are not so easy. Like existentialism, the movement referred to here as "coexistentialism" resists definitions and "What is . . . " questions that seek to identify it. An existentialism defined is an existentialism denied. The same holds, *mutatis mutandis*, for coexistentialism. If you identify coexistentialism, you lose it. It is not unlike what Augustine says of the incomprehensibility of God: If you understand it, it is not God (*si comprehendis, non est deus*).[25] *Si comprehendis non est coexistentialismus.* Without defining coexistentialism or attempting to explain or understand it, one can nonetheless begin to practice it, live it, get a sense of it. In the context of introductory remarks, a consideration of some of the first figures to introduce the idea of coexistentialism can convey a sense of coexistentialism, gesturing toward coexistentialism without giving a proper definition.

Gestures toward coexistentialism appear throughout existentialism. Although they are different, there is no clear and distinct boundary between them. Karl Jaspers (1883–1969), in his articulation of a philosophy of existence (*Existenzphilosophie*), gestured toward the transition from existentialism to coexistentialism. The transition has to do with overcoming loneliness. "Existence-philosophy is," according to Jaspers, "the way of thought by means of which man seeks to become himself; it makes use of expert knowledge while at the same time going beyond it."[26] That fits standard definitions of existentialism, including the gender exclusive language.[27] For Jaspers, the philosophical situation of *Existenzphilosophie* has its beginnings in Kierkegaard and Nietzsche. Although Kierkegaard and Nietzsche are quite different in some respects, they share some suspicions toward authority, critiques of institutions, affirmations of freedom, and alas, loneliness. Their emphasis on the exceptional existence of every individual was uncommon, indeed, exceptional. Neither man found public acclaim in his life, but each opened a way for new forms of intimacy and public coexistence. "Although without position, marriage, without any effective role in existence, they nevertheless appear as the great realists, who had an authentic feeling for the depths of reality."[28]

The lonely lives of Kierkegaard and Nietzsche mark the beginning of a new form of humanity that can now be shared, communicated, distributed, ecologized. In that sense, Kierkegaard and Nietzsche, "having dared to be shipwrecked," are like "representative destinies, sacrifices," or "modern martyrs," whose lonely existentialism opened up a horizon for coexistentialism.[29] Of course, Kierkegaard and Nietzsche would not conform to typical conceptions of destiny, sacrifice, and martyrdom, but that is exactly why those words do apply to them. Each of them personally lived through a shipwreck, a breakdown of the meaning and grounds of human existence, including a breakdown of categories like destiny, sacrifice, and martyrdom. Living a destiny without destiny as martyrs after martyrdom, they have prepared and continue to prepare future generations for the ongoing breakdown of the conditions of human existence, which is ultimately a breakdown of the conditions of life on Earth. Their works prepare humans to take up the work of engaging real beings and communicating the truth about the actual conditions of our existence. Truth is not some private discovery or individual experience. The truth of existence takes place through communication, not exclusively human communicative action but a communication of being, such that being itself is shared interpretively, "reality-through-communication."[30] The lonely beginnings of existentialism prepared the way for a communicable existentialism, a shared existentialism.

The *Existenzphilosophie* of Jaspers gestures toward coexistentialism, but it is still relatively anthropocentric compared with the ecological orientation of coexistentialism. Jaspers expresses a much more hospitable attitude toward science than some other existentialist critiques and revolts against science, not least of which is the underground man's refusal to accept that two and two is four. However, Jaspers was primarily concerned with human existence, not ecological coexistence. A similar shift toward coexistentialism can be seen in the work of Martin Heidegger. Heidegger defines human existence in terms of being-with (*Mitsein*) and being-in-the-world, and his writings have been influential for numerous ecological thinkers. Heidegger's thinking dismantles the subject-object dichotomy that sets humans as subjects over against the world as a totality of mere objects or resources. His work makes significant strides on the path to welcoming the hiddenness, withdrawal, strangeness, and difference that have been suppressed or excluded throughout the history of philosophy. However, his criticisms of science make it difficult to take ecological science as a serious matter for thinking, and all of his talk about being-with and dwelling still presupposes the solitude of the existing individual. Furthermore, Heidegger tends to reduce nonhumans to a privative status in relation to humans, as in his famous distinction between "world-forming" (*weltbilden*) humans, "poor in world" (*weltarm*) animals, and "worldless" (*weltlos*) stones.[31] A special kind of exclusion is reserved for plants, which have no place in Heidegger's schema, not even poor, not even worldless. This calls for rectification.[32]

Coexistentialism starts showing up more explicitly in post-Heideggerian thinkers, that is to say, thinkers who follow Heidegger in treading an unbeaten path away from the isolated subject and toward a mode of existence that is open to the world, in intimate contact with beings, dwelling with strangeness and difference. For example, in Morton's proposal for the development of "an ethical attitude we might call 'coexistentialism,'" he notes Emmanuel Levinas and Luce Irigaray as two examples of "prominent recent theorists of 'coexistentialism.'"[33] Levinas and Irigaray are both thinkers of otherness and difference. Thinking in the wake of Heidegger while also diverging from Heideggerian philosophy in critical ways, they move further away from loneliness and toward an affirmation of vulnerability and intimacy, which for Levinas takes the form of a phenomenological description of ethical encounters with others, and for Irigaray it takes the form of a feminist theory of sexual difference.

Along similar lines, in the "ecological existentialism" articulated by Deborah Bird Rose, she finds feminist theory and the existential phenomenology of Levinas to be crucial sources. She does not refer to Irigaray though. The main feminist theorists to which Rose refers are Val Plumwood and Donna Haraway, who are very different from Irigaray in some respects, yet they share commitments to relationships that do not efface the differences with which humans are woven together with one another and with all of the elements of existence.[34] In addition, ecofeminism and phenomenology also figure prominently in Andy Fisher's approach to ecopsychology, which he describes as an "ecological existentialism," which for him means that "the ultimate concerns of life are worked out in the context of our membership within the community of all life."[35]

The perspicacity of phenomenological description along with the critical and constructive force of feminist theory provide coexistentialism with ways of attending to the exigencies of real bodies, getting in touch with things in all of their messiness, from their finely textured details to their most astonishing, horrifying, and infinitely compelling adumbrations. As Levinas points out, there is an instability to things that renders them open and vulnerable to their elemental environment, in the way that wood can disappear into smoke and ashes. "A thing exists in the midst of its wastes."[36] There is an "extreme fragility" to things, a sort of "exorbitant ultramateriality" or "paroxysm of materiality" whereby things are exposed to what lies outside their own limits, an "exhibitionist nudity" whereby things expose themselves to an "unreality at the threshold of the real."[37] As Judith Butler points out, understanding vulnerability is part of understanding the dynamics of social and political life, including wounds of oppression and violence. "The body implies mortality, vulnerability, agency: the skin and the flesh expose us to the gaze of others but also to touch and to violence," and a different conception of political coexistence could open up, perhaps, "by staying with the thought of corporeal vulnerability itself."[38]

Coexistentialism can foster a democratic politics of care for all beings folded into the mesh of ecological coexistence. It is in this sense that the term "coexistentialism" is deployed by Ella Myers.

> Coexistentialism refers to an ecological perspective that takes the interconnectedness of all worldly entities, humans and nonhumans, organic and inorganic matter, as its starting point. . . . Coexistentialism locates human beings within this worldly "mesh" rather than above it. Such an ecological awareness challenges fantasies of mastery by reminding people of the extent to which they are affected by the doings of nonhumans. . . . A coexistentialist perspective can and should guide the way we care for the world, as both home and in-between."[39]

Myers takes the term "coexistentialism" from Morton, whose use of the term conveys an insight into the "radical intimacy" of existence, which is to say, the insight that "existence is always co-existence."[40] Myers notes that the outlook of coexistentialism is also shared by Jane Bennett and Bruno Latour, although they do not use the term.[41] In different ways, Latour and Bennett each develop theories that account for the intimate interconnectedness of humans and nonhuman agents ("actants") in complex networks ("assemblages").

> Latour and Bennett believe that seeing and thinking in terms of actants and assemblages is a necessary corollary to acknowledging interdependence between humans and nonhumans. They maintain that the acknowledgment of coexistentialism—deep interconnectedness between all living and nonliving things—requires letting go of conventional notions of agency. Attributing effects to assemblages rather than to human subjects, they claim, honors the basic truth of coexistence while also undermining dangerous norms of instrumentality and fantasies of mastery. They contend that if humans see actants all around them, they will be less inclined to assume a controlling posture toward nonhuman entities and more interested in nourishing the relational webs and ecosystems in which they participate.[42]

Latour and Bennett do not use the terms "coexistentialism" or "existentialism" to describe their works. Latour's theoretical approach is often described as "actor-network theory," although more recently he has expanded that model to include not just actors and networks but the diverse "modes of existence" in which actors and networks circulate.[43] Bennett's approach is one among others associated with "new materialism," which differs from old, traditional materialism insofar as it does not view matter as passive, inert substance but as vibrant, dynamic agency or "thing-power," like the strange vitality of Kafka's spool of thread, Odradek.[44] In their attempts to articulate the agencies of actors and the powers of things, both Latour's and Bennett's approaches overlap with Morton's philosophy, which he describes as "object-oriented ontology"—"a unique form of realism and nonanthropocentric thinking."[45] Another proponent of object-oriented ontology, Levi Bryant, likewise proposes an "existential ecology."[46]

Of course, from an existentialist perspective, discerning differences is important, but hashing out the contrasts and comparisons between different labels, "isms," and approaches can become a tedious distraction, like focusing your attention on the details of a menu and forgetting about the actual meal, or trying to eat your chopsticks instead of your noodles. The point here is not to argue about the territories staked out by different schools of thought. Rather, the point is to introduce this coexistentialism which is not one.[47] In other words, the point is to indicate that there are many ways to engage in coexistentialism, including ways that do not use explicitly existentialist labels like "coexistentialism" or "ecological existentialism." Those terms are neither necessary nor sufficient conditions for coexistentialism. Using those terms could even preclude an affirmation of coexistence, resulting instead in an affirmation of a terminological convention.

The history of the term "coexistentialism" is itself quite unremarkable, in contrast to the controversy surrounding various applications and disavowals of the label "existentialism." The most prominent existentialists—Kierkegaard, Dostoevsky, Nietzsche, Heidegger, Sartre, Camus—

all developed or at least began developing their thought before Gabriel Marcel coined the term "existentialism" in 1943.[48] References to existential phenomenology, *Existenzphilosophie,* and existential truth had been circulating much earlier, and when the "ism" emerged, many of the thinkers to whom it supposedly applied rejected it. In contrast, the term "coexistentialism" appears around the same time that coexistentialists like Levinas and Irigaray came to prominence. Some of Levinas' most famous works (e.g., *Totality and Infinity*) were written in the 1940s, the term "coexistentialism" was first coined in the following decade, and Irigaray gained notoriety with her first works in the 1970s, such as *This Sex Which Is Not One.*

Gaston Bachelard seems to have first coined the term "coexistentialism" in 1958 in his phenomenological account of images of intimate space, *The Poetics of Space.* Bachelard uses the term "coexistentialism" in his observation that human "consciousness of our own existence" participates intimately in the space of the "coexistence of things": "In this coexistentialism every object invested with intimate space becomes the center of all space. For each object, distance is the present, the horizon exists as much as the center."[49] That sense of coexistentialism is not definitive for all coexistentialists, but it is indicative of some of the shared topics addressed by coexistentialists, including the intimacy of human and nonhuman entities and the implosion of foregrounds and backgrounds (e.g., the center and the horizon; the organism and the environment; the self and the other; the familiar and the uncanny).

Subsequent to Bachelard's usage, the term "coexistentialism" gained relatively little attention, at least compared to the controversy that erupted with the coining of "existentialism" fifteen years earlier. Coexistentialism did not then become a trend or an object of disavowal or disdain. In opposite fashion to the tendency of most existentialists to dislike the label "existentialism," most coexistentialists have not heard the term "coexistentialism," or if they have heard it, they do not write about it. For instance, while Morton uses the term, he considers Levinas and Irigaray to be coexistentialists, although they never use any such term. Similarly, Alphonso Lingis, would surely be considered a coexistentialist. A translator of existential phenomenologists like Levinas and Maurice Merleau-Ponty, Lingis is even more carefully attentive to the compelling allure of nonhumans than were his predecessors. As Lingis says, "I am a pulse of care in the universe."[50] That care extends to all things, humans and nonhumans, as they issue imperatives to attend to their unfathomable mysteries.

> A thing is before us as a task and a reality embedded in the subsistence of the world inasmuch as it is closed in itself and harbors secrets and surprises. The world-order which sustains it, the levels which maintain

> it, are presented not as a framework surveyed but as directives pursued.[51]

Along with Lingis, there are many others who have extended existential philosophy into the radical intimacy of coexistence, although they do not use the terms "coexistentialism" or "ecological existentialism." For example, Jean-Luc Nancy and Peter Sloterdijk are two of the most prominent contemporary coexistentialists drawing deeply from Heidegger's philosophy. Elaborating on Heideggerian notions of inhabiting a world and being-with, Nancy and Sloterdijk extend those notions outside of the anthropocentric and provincial tendencies in Heidegger's thought. Sloterdijk's massive *Spheres* trilogy is oriented toward a recuperation of the intimate space characteristic of being-in-the-world and a refutation of the loneliness (*Einsamkeit*) that Heidegger considers fundamental to inhabiting a world.[52] If Sloterdijk placed a sign at the entrance of the *Spheres* trilogy, this would be its message: "let no one enter who is unwilling to praise transference and to refute loneliness."[53] Nancy's works articulate a sense of the withness (the "co-") of coexistence, such that "we" does not refer exclusively to humans but to a differentiated and entangled plurality of beings, such that "we" can affirm the struggle to "*say we for all being, that is, for every being, for all beings one by one, each time in the singular of their essential plural.*"[54]

In sum, coexistentialism is not derived from one thinker or one school of thought. Although there are people who have contributed much to the transition toward an existentialism of intimate coexistence and ecological interconnectedness, coexistentialism does not have a founder. The notion of a founder or foundation connotes the sort of stable background that coexistentialism finds untenable. Some figures have contributed more to coexistentialism than others, but there is no rigid boundary separating precursors from founders or separating founders from followers, and there is no firm line separating those who are in the movement to those who are outside of it. Along with thinking of coexistentialism in terms of the contributions of particular theorists, one can also think of coexistentialism in terms of the fields of study with which it is closely affiliated, including new materialism, object-oriented ontology, existential phenomenology, and feminist theory. Furthermore, the emerging interdisciplinary field of ecological or environmental humanities has been the locus for much coexistentialist work. Hence, alongside phenomenology and feminism, ecophenomenology and ecofeminism have played significant roles in facilitating the transition to coexistentialism.

Another area of environmental humanities that is important in this regard is ecocriticism. Sometimes known as ecopoetics or green cultural studies, ecocriticism brings the theories and methods of literary criticism into contact with ecological perspectives. This means giving ecological readings of literary texts, which would not only provide an ecological perspective on a text like John Milton's *Paradise Lost* but would also consider how Milton's work itself already has an ecological perspective *avant la lettre*.[55] Furthermore, ecocriticism means more than just adding ecological ingredients to literary criticism and stirring. In its more radical instantiations, it also implies that the truth about ecology—the study of relationships between things (organisms and environmental conditions)—involves something like a literary dimension, such that aesthetics, interpretation, and metaphor characterize all ecological interactions and are not just phenomena relegated to human consciousness.[56] In this sense, ecocriticism does not just analyze stories about Earth or bring a semiotic perspective to matter. Its investigations indicate that Earth is itself storied, and materiality is semiotic. For instance, instead of analyzing poems about animals, ecocriticism can analyze the poetry that animals make themselves.[57] Ecological realities are literary, not simply literal. This ecocritical point bears upon the larger context of environmental humanities. Like its close neighbor, posthumanities, theories and practices of environmental humanities indicate that inquiry in the humanities is not just for or about humans.[58] Indeed, it is not even entirely by humans, as any human author is constituted in multispecies networks of human and nonhuman agents. For scholars of environmental humanities, meaning and semiotics is not exclusively human property but is distributed across ecosystems.[59]

The explicit orientation toward ecology, including scientific ecological perspectives, sets coexistentialism distinctly apart from existentialism. Aside from that, coexistentialism and existentialism share many similar affiliations with other schools of thought. They both connect with phenomenology, feminism, and various "post-" movements, such as posthumanism, postcolonialism, postmodernism, and poststructuralism. They share close contact with critiques of capitalism in Marx and critical theorists, although affiliations with leftist politics are more common among coexistentialists, as there are also some conservative tendencies in existentialists, challenging the possibility of democracy while frequently bemoaning the inauthenticity of the masses and their crude leveling of the singularity of existing individuals. As John Caputo observes, "Kierkegaard, Nietzsche and Heidegger were politically conservative, even ultraconservative, thinkers, alerting us to the downside and the dangers posed by democratic institutions."[60] Indeed, in its most nihilistic expressions, existentialism can become completely apolitical, seeing nothing but the downside of all worldly affairs. Hans Jonas observes this in his comparison between nihilistic tendencies in existentialism and in the ancient philosophical spirituality of Gnosticism, specifically the tendency to posit a dualism between a meaning-making human and an evil cosmos (Gnosticism) or a cosmos so indifferent that it is not even evil (existentialism).[61] For coexistentialists, even a conservative thinker like Sloterdijk still embraces some aspects of communism in what he calls "co-immunism," and he is highly critical of the individualist politics foaming all over the planet.[62] Existentialism and coexistentialism both resonate with nihilism, sharing some irreverent and rebellious tendencies and affirming the collapse of the metaphysical systems of the classical and modern worlds, but most of the thinkers associated with existentialism and especially coexistentialism affirm the existence of real things or events, thus looking like pulses of care in the world, not nihilists at all. Maybe there is no "world," and maybe beings never completely "be," but that uncertain possibility is evidence that something is happening, something matters. Maybe it is not really something, but it is not nothing either.

To imagine the difference between existentialism and coexistentialism, consider the following three figures. First, the underground man in Dostoevsky's writings is underground in the sense of isolation and low social status, whereas the notes from the coexistentialist underground convey many kinds of subterranean realities, including phenomena like alienation and injustice, subliminal meaning and subtext, the countercultural underground, as well as connections with actual subterranean bodies—roots, rhizomes, fossils, fossil fuels, worms, toxic wastes, earthquakes, magma, time capsules, aquifers, and wine cellars.[63] Most subterranean of all is the unfathomably deep reality of beings, their irreducible otherness and strangeness. Second, for Camus, the basic philosophical question is how to affirm one's existence and face the possibility of suicide, and for coexistentialists one could suggest that the basic philosophical question is how to affirm coexistence in a time of mass extinction, not only facing one's possible suicide but also facing irrevocably actual biocide. Third, one of Sartre's well-known works declares that existentialism is a humanism, and in response, one could argue that coexistentialism is a posthumanism, although one should be cautious before committing to posthumanism or any other postal tendencies.[64] Those three figures—the underground, extinction, and posthumanism—are not intended as a means for closing or concluding this section but for opening out, as an attempt or a temptation to begin fashioning coexistentialist outfits.

CHAPTERS

Extending the overview of existentialism given in the present chapter, the second chapter elaborates on five of the enduring legacies of existentialism. First and foremost is style. Existentialism is a style of writing and thinking that affirms indirect and open-ended forms of expression, a freestyle where there is no theoretical closure or rigid "-isms" (including existential*ism* itself). Another legacy is a relentless affirmation of experience in the predominantly programmatic, formulaic, and institutionalized structures that confine experience. The third is resistance, including resistance to transcendent or institutional forms of authority, resistance to conformity and convention, and resistance to meaninglessness. A fourth legacy is a postsecular orientation, which maintains resistant yet allied relationships with religious traditions and their secular humanist counterparts, overcoming the sacred-secular distinction in the unheard-of and detuned (*absurdus*) resonance of the absurd, which is the fifth legacy. Existentialism affirms nonrational ways of knowing, celebrating paradox, and embracing uncertainty, impossibility, *aporia*, and altered states of consciousness. Subsequent chapters discuss these five legacies as they are taken up in various ways by coexistentialists like Rose, Morton, and many other contemporary theorists who do not necessarily call themselves coexistentialists or ecological existentialists but exhibit shared commitments and characteristics with those who do (e.g., Sloterdijk, Nancy, Irigaray, Haraway).

The third chapter focuses on the situation of coexistentialists after God and nature, including two senses of "after." First, it is subsequent to God, nature, and all totalizing categories that render the world into two clear and distinct categories of transcendence and immanence or intelligibility and sensibility. The demands of planetary coexistence in a state of emergency have rendered those categories untenable, as indicated by numerous announcements of the deaths of God and nature. Coexistentialism orients itself to religion and ecology without God and nature. Second, coexistentialism is after God and nature in the sense of seeking or tracking God and nature, following God and nature as they slip outside of inherited dualisms of natural/cultural, material/divine, physical/spiritual, etc. Being after God and nature in both of those senses, coexistentialism opens up possibilities for an updated form of animism—a postsecular animism for our planetary era, an animism without animism. Such animism is not unlike the "new animism" proposed by Graham Harvey and other anthropologists studying indigenous traditions. However, animism without animism does not require adherence to indigenous worldviews, nor does it require adherence to anthropological prejudices and preconceptions. A Christian, atheist, Buddhist, or Muslim can also participate in this animism without animism insofar as they resist the assimilation of ontological diversity into dualistic categories (subject/object, transcendent/immanent, human/nonhuman, animate/inanimate). This is a strange animism, as animistic agencies today are showing up in unlikely and rather horrifying phenomena, such as toxic waste, nuclear radiation, climate change, plastic, Styrofoam, and other dangerous newcomers to planetary coexistence.

The next chapter, "Remaining Exposed," explores the role of vulnerability in coexistentialist affirmations of intimacy. Vulnerability discloses an aporia of coexistence, as it allows you to exist by making you open to destruction. It renders things possible and impossible. Haraway, Sloterdijk, and Nancy are the three coexistentialists most engaged in questions of exposure; that is, they are explicitly concerned with the challenges of remaining exposed to others. For Nancy, remaining exposed is evident in his reflections on his heart transplant. Exposure makes precarious contact with human and nonhuman beings in a way that touches without touching. For Sloterdijk, remaining exposed is a way of participating in the construction of shared intimate space in close personal relationship and in the collective scale of global civilization. Haraway provides an example of a practice of remaining exposed in her practice of companioning with her dog, a practice of touching for which becoming-with is also becoming worldly. Sloterdijk, Nancy, and Haraway each open and extend Heidegger's sense of exposure in the open, the ecstatic exposure of *Dasein*, abandoning Heidegger's relatively provincial, anthropocentric interpretation of being-with (*Mitsein*). The intimacy of withness is not restricted to humans but pervades all beings. Existence is always already coexistence.

The fifth chapter elucidates coexistentialist understandings of spatiality in terms of "Roundness." Just as the previous chapter radicalized Heidegger's anthropocentric conception of being-with, this chapter radicalizes his conception of the inhood (*Inheit*) of being-in-the-world. Whereas Heidegger's articulations of space and place are lacking in specificity and diversity, Gaston Bachelard's coexistential analysis of spatial imagination criticizes Heidegger's abstractness and provides a more detailed account of the numerous ways that humans enter into intimate relations with the curvature of space—the roundness of things—ranging from the miniature to the cosmically immense. Peter Sloterdijk's monumental *Spheres* trilogy expands on Heidegger's notion of being-in and Bachelard's notion of roundness even further, giving a comprehensive history—indeed, a grand narrative—of intimate spaces (i.e., spheres), showing how contemporary ecological and social crises of globalization are related to a modern individualist immunity to intimacy. I discuss place in a glance to consider how contemplative ecological practices can facilitate the construction of intimate spaces.

Notes from the coexistentialist underground are musical notes, attuning to the strange tunes of other beings entangled in ecological coexistence. Accordingly, this book includes a brief "Interlude" at the middle of the work, introducing a coexistentialist sense of sustainability in terms of liminality, the in-betweens, intermezzos, and intervals that connect *and* differentiate humans and nonhumans. This sense of liminality can be understood as a generalized theory of media, a theory of intermediaries (Sloterdijk's spheres), or a theory of what Gilles Deleuze and Félix Guattari call "interbeing" (a term also used by the Zen Buddhist monk Thich Nhat Hanh). This is an updated version of the golden mean of Aristotelian and Confucian ethics, updated because ecological emergency complicates those classical attempts to discern hierarchies of extremes and means, peripheries and centers. Coexistentialist liminality is uncertain, ambiguous, and strange, without any determinate norms and prescriptions.

The next chapter, "After Humanism," considers what comes after the human of humanism. Whereas existentialists tend to hold a humanistic view for which humans are the only source of meaning in an otherwise meaningless world, coexistentialists tend to favor a more posthumanistic view, for which humans are always already entangled with nonhuman actors, and all matter is intrinsically semiotic or meaningful. However, going postal is not necessarily the best move. What is called for is to become human again and still for the first time, which means becoming a human engulfed in multitudes of nonhumans. It means becoming anthropocosmic, in a sense laid out by Bachelard. Phenomenology, poetry, and religions provide techniques of anthropocosmic ecstasy that can facilitate practices of becoming human, as long as those practices are updated for an era in which humans are intimately enmeshed with strange newcomers, entities like global capitalism, mass extinction, the Internet, plastic, and global warming, entities that Morton calls "hyperobjects." In an age of hyperobjects, being a human means affirming powerlessness in order to become possible, which is to say, it means being what Morton calls with Dominic Boyer a "hyposubject." This opens up new possibilities for the humanities to become ecological humanities.

The eighth chapter builds on the previous chapter's articulation of hyposubjects—humans becoming anthropocosmic amid ecological emergency, meshing with swarming singularities of nonhumans. This chapter discusses the aesthetic and ethical implications of human existence in terms of "looking good," a concept that the feminist philosopher Elizabeth Grosz develops in reference to the challenge of living life like a work of art, looking well and being looked at well, looking good at and for the other. Looking good, hyposubjects multiply and intensify differences. Although it was mostly overlooked by existentialists, sexual difference has a particularly important role to play in coexistentialism, particularly in light of the work of Irigaray and, more recently, Catherine Malabou. An understanding and affirmation of sexual difference supports understandings and affirmations of all differences, including ontological difference. Furthermore, as Grosz demonstrates, Darwin's concept of sexual selection is compatible with a feminist theory of sexual difference. This extended concept of sexual differences contributes to an understanding of the differences at stake in current emancipatory politics, including differences of race, class, gender, sex, and species, and it also contributes to a more profound understanding of the role of drugs in facilitating altered states of consciousness. I consider psychedelic drugs as a catalyst or amplifier of sexual difference. In short, this chapter indicates that hyposubjects are feminist, queer, prismatic, ecological, and psychedelic.

After God and nature, after humanism, coexistentialism is also after the end of the world. There is no longer a world in the sense of a background or a container in which things take place. There is no established world or given totality of all things. The ninth chapter ("Becoming Worldly") elucidates the object-oriented ontology by which Morton elucidates the ecology of things that remains after the end of the world, which is an ecology of nothingness or emptiness. This perspective is at odds with the concept of worlding embraced by Rose, Haraway, and new materialists for whom worlding is a complex, dynamic process of interrelationship, of which specific things are a product. If there is worlding after the end of the world, it is not as an underlying process but, following a deconstruction conveyed by Nancy as well as Jacques Derrida, as the sheer coexistence of beings, a crowd of intimate alterities, each one of which is *the* world, singularly and irreplaceably, and each one of which I must bear.

The phrase "shut up and train" is one of the slogans that Haraway proposes to represent the becoming-with and becoming worldly of companion species. The tenth chapter adopts this slogan, focusing on the role of training or exercise (*askesis*) in the postsecular ecology of coexistentialism, particularly in light of Sloterdijk's call for cooperative asceticisms to build a planetary immune system that supports intimate coexistence and wards of rapacious exploitation and individualist loneliness. Haraway and Sloterdijk both affirm the importance of cooperative training, but Haraway is more emphatic than Sloterdijk that training must be a multispecies effort. As an example, I discuss coexistential exercises in listening based on the listening practices expressed by Nancy and the American forester and conservationist Aldo Leopold.

The eleventh chapter discusses the concept of the axial age and its relevance for understanding existentialism and the transition represented by the emergence of coexistentialism. Karl Jaspers coined the term "axial age" (*Achsenzeit*) to describe the period of history (around 2,500 years ago) when philosophy, science, and classical religions first emerged, articulating a conception of universal law and universality humanity, exemplified in paradigmatic individuals like Socrates, the Buddha, and Confucius. Jaspers saw Kierkegaard and Nietzsche as indications that the ways of life opened up in the axial age are exhausted. Kierkegaard and Nietzsche shipwrecked with God, nature, humans, the world, carrying and being carried away by the end of the axial world. Are we arriving at a second axial age, a new age that will be formative for millennia to come? Is coexistentialism a New Age movement? Not quite. If there is another axial age, it must be an axial age without an axis, a historical turning point where humans give up trying to fit together their own timeline and locate their turning points. In ecological emergency, humans do not have the time. Whatever age this is, or whatever ages these are, a radical asymmetry interrupts the time-scales of almost 7.5 billion humans with the times of trillions upon trillions of nonhumans.

The concluding coda provides some summary remarks, briefly reiterating some of the main themes of coexistentialism introduced in the book and, however hesitantly, mentioning something personal about my way into coexistentialism, not by way of a litany of categories that situate my subject position, but by way of something more intimate, a personal tale for a literary "tail" (*coda*). Regarding the specific lines of difference that cross my subject position, I am a white, heterosexual, male, human, in my mid-30s, from the southwestern United States, living in the San Francisco Bay Area, working as an adjunct professor. I do not have disabilities, but I am not exactly neurotypical either. Perhaps "coexistentialist" should be added to the list, if only to indicate that I do not take those categories or any categories for granted. I am horrified and fascinated by my subject position and my relations with others—human, nonhuman, and otherwise. Sincerely, I acknowledge my hypocrisy, my complicities and privileges as well as my powerlessness and possibility. I am not here to peddle an "ism" or solve any problems. I am here to get in the fray with you, for you, for us, whoever we are—shimmering singularities, intimate alterities. Refuting cynicism and loneliness, I am attempting to speak to you and to us, to we who may be. We are here, with more to come.

NOTES

1. Jean-Paul Sartre, *Being and Nothingness: A Phenomenological Essay on Ontology*, trans. Hazel E. Barnes (New York: Washington Square Press, 1984); Walter Kaufmann, ed., *Existentialism: From Dostoevsky to Sartre*, rev. and exp. ed. (New York: Penguin Books, 1988).
2. Hunter S. Thompson, "A Man Has to Be Something, He Has to Matter: Hunter S. Thompson to Hume Logan," in *Letters of Note: An Eclectic Collection of Correspondence Deserving of a Wider Audience*, ed. Shaun Usher (San Francisco: Chronicle Books, 2014), 65.
3. Deborah Bird Rose, *Wild Dog Dreaming: Love and Extinction* (Charlottesville: University of Virginia Press, 2011), 42.
4. Björk, "Jóga," *Homogenic* (London: One Little Indian, 1997). The attribution of "best eco anthem" to this song comes from Timothy Morton in a post from his blog. Timothy Morton, "The Best Eco Anthem," *Ecology without Nature* blog, October 29, 2014, ecologywithoutnature.blogspot.com/2014/10/the-best-eco-anthem.html. A creative and mutually appreciative dialogue between Morton and Björk is included in the volumes that accompanied Bjork's show at the Museum of Modern Art, Björk, *Björk: Archives* (London: Thames and Hudson, 2015).
5. Timothy Morton, "Deconstruction and/as Ecology," in *The Oxford Handbook of Ecocriticism*, ed. Greg Garrard (Oxford: Oxford University Press, 2014), 301.
6. Friedrich Nietzsche, *Twilight of the Idols*, trans. Richard Polt (Indianapolis: Hackett Publishing, 1997), 9.
7. Søren Kierkegaard, *Papers and Journals: A Selection*, trans. Alastair Hannay (London: Penguin Books, 1996), 212.
8. Fyodor Dostoevsky, *Notes from Underground*, trans. Richard Pevear and Larissa Volokhonsky (New York: Alfred A. Knopf, 2004), 14.
9. Ibid., 30.
10. Ibid., 32.

11. In William Barrett's well-known introduction to existentialist philosophy, he describes the existentialist as an *Irrational Man*, which would be a more apt description without the gender exclusive language. The work was originally published in 1958, before gender inclusive language was a common practice in academia. Nonetheless, beyond adherence to linguistic conventions, there are androcentric and patriarchal tendencies in existentialism (see chapter 8 of this book). In any case, Barrett's work remains a standard representation of the irrational or nonrational elements of existentialism. William Barrett, *Irrational Man: A Study in Existential Philosophy* (New York: Anchor Books, 1990).

12. Søren Kierkegaard, *Philosophical Fragments*, trans. Howard Hong and Edna Hong (Princeton, NJ: Princeton University Press, 1985), 37.

13. Simone de Beauvoir, *The Ethics of Ambiguity*, trans. Bernard Frechtmen (New York: Citadel Press, 2000).

14. Timothy Morton, *The Ecological Thought* (Cambridge: Harvard University Press, 2010), 8.

15. Albert Camus, *The Myth of Sisyphus and Other Essays*, trans. Justin O'Brien (New York: Vintage Books, 1991), 3.

16. Camus, *The Myth of Sisyphus and Other Essays*, 122.

17. Ibid., 119.

18. Camus, *The Myth of Sisyphus and Other Essays*, 123.

19. The contested contributions to ecological thought and practice in the works of Nietzsche, Heidegger, and other philosophers associated with postmodernism are subject to ongoing debate. Published over twenty years ago, Michael Zimmerman's book on this topic is still one of the most thorough and comprehensive. Michael E. Zimmerman, *Contesting Earth's Future: Radical Ecology and Postmodernity* (Berkeley and Los Angeles: University of California Press, 1994).

20. Timothy Morton, *Ecology without Nature: Rethinking Environmental Aesthetics* (Cambridge: Harvard University Press, 2007).

21. Jane Bennett, *Vibrant Matter: A Political Ecology of Things* (Durham: Duke University Press, 2010), 7. In "The Cares of a Family Man," Kafka presents this description of Odradek: "It looks like a flat star-shaped spool of thread, and indeed it does seem to have thread wound upon it; to be sure, these are only old, broken-off bits of thread, knotted and tangled together, of the most varied sorts and colors . . . the whole thing looks senseless enough, but in its own way perfectly finished." Franz Kafka, "The Cares of a Family Man," in *The Complete Stories*, ed. Nahum N. Glatzer, trans. Willa and Edwin Muir (New York: Schocken, 1971), 428.

22. Robert Brody, "The Existential Genius of Late Woody Allen," *The New Yorker*, July 16, 2015, www.newyorker.com/culture/cultural-comment/the-existential-genius-of-late-woody-allen.

23. A brief overview of the philosophical implications of the film is articulated in a review by John Snider, "I Heart Huckabees," *Philosophy Now: A Magazine of Ideas* 51, June/July 2005, philosophynow.org/issues/51/I_Heart_Huckabees.

24. *Curb Your Enthusiasm*, "Beloved Aunt," *Curb Your Enthusiasm Episode Scripts*, www.springfieldspringfield.co.uk/view_episode_scripts.php?tv-show=curb-your-enthusiasm&episode=s01e08. The line from Beckett is, "you must go on, I can't go on, I'll go on." Samuel Beckett, *The Unnamable* (London: Faber and Faber, 2012), 134.

25. Augustine, *Sermo 117*, quoted in Jean-Luc Marion, *In Excess: Studies of Saturated Phenomena*, trans. Robyn Horner and Vincent Berraud (New York: Fordham University Press, 2002), 155.

26. Karl Jaspers, *Man in the Modern Age*, trans. Eden Paul and Cedar Paul (New York: Routledge, 2010), 159.

27. On gender and sexual difference in existentialism and coexistentialism, see chapter 8 ("Looking Good").

28. Karl Jaspers, *Reason and Existenz: Five Lectures*, trans. William Earle (Milwaukee: Marquette University, 1996), 29.

29. Ibid., 38.

30. Ibid., 79.

31. Martin Heidegger, *The Fundamental Concepts of Metaphysics: World, Finitude, Solitude*, trans. William McNeill and Nicholas Walker (Bloomington: Indiana University Press, 1995), 196.

32. The works of Michael Marder and Matthew Hall are exemplary inquiries into the place of plants and vegetal life in the history of philosophy. Hall's work is particularly useful for considering indigenous and Asian perspectives as well as Western philosophical and religious perspectives. Matthew Hall, *Plants as Persons: Toward a Philosophical Botany* (Albany: SUNY Press, 2011). Marder brings phenomenology, hermeneutics, and deconstruction together in the interest of articulating a vegetal philosophy. Michael Marder, *Plant-Thinking: A Philosophy of Vegetal Life* (New York: Columbia University Press, 2013). For a compelling account of plants in the history of the European philosophical tradition, see Michael Marder, *The Philosopher's Plant: An Intellectual Herbarium* (New York: Columbia University Press, 2014).

33. Morton, *The Ecological Thought*, 47, 144n96. On Levinas' connection to coexistentialism along with other Jewish existentialists like Martin Buber, Franz Kafka, and Hans Jonas, see Paul Mendes Flohr, "Jewish Co-Existentialism: Being with the Other," in *Situating Existentialism: Key Texts in Context*, ed. Jonathan Judaken and Robert Bernasconi (New York: Columbia University Press, 2012).

34. Rose, *Wild Dog Dreaming*, 15, 112.

35. Andy Fisher, *Radical Ecopsychology: Psychology in the Service of Life* (Albany: SUNY Press, 2002), 267n9.

36. Emmanuel Levinas, *Totality and Infinity: An Essay on Exteriority*, trans. Alphonso Lingis (Dordrecht: Kluwer Academic Publishers, 1991), 139.

37. Ibid., 256.

38. Judith Butler, *Undoing Gender* (New York: Routledge, 2004), 21, 23.

39. Ella Myers, *Worldly Ethics: Democratic Politics and Care for the World* (Durham, NC: Duke University Press, 2013), 127, 130.

40. Ibid., 180n55. Morton, *The Ecological Thought*, 47.

41. Myers, *Worldly Ethics*, 180n55.

42. Ibid., 101.

43. Bruno Latour, *An Inquiry into Modes of Existence: An Anthropology of the Moderns*, trans. Catherine Porter (Cambridge: Harvard University Press, 2013).

44. Bennett, *Vibrant Matter*, 6.

45. Timothy Morton, *Hyperobjects: Philosophy and Ecology after the End of the World* (Minneapolis: University of Minnesota Press, 2013), 2.

46. Levi Bryant, "Black," in *Prismatic Ecology: Ecotheory Beyond Green*, ed. Jeffrey Jerome Cohen (Minneapolis: University of Minnesota Press, 2013), 292, 301.

47. This is an allusion to Luce Irigaray, *This Sex Which Is Not One*, trans. Catherine Porter (Ithaca: Cornell University Press, 1985).

48. For an overview of the history of existentialism, including the coining of the term "existentialism" and the emergence of various existence-oriented philosophies, see Christine Daigle, ed., *Existentialist Thinkers and Ethics* (Montreal: McGill-Queen's University Press, 2006), 5ff.

49. Gaston Bachelard, *The Poetics of Space*, trans. Maria Jolas (Boston: Beacon Press, 1994), 203.

50. Alphonso Lingis, *Sensation: Intelligibility in Sensibility* (Atlantic Highlands: Humanities Press, 1996), 83.

51. Ibid., 35.

52. The "solitude" of Heidegger's philosophy, such as presented in his 1929–1930 Freiburg lecture, *Fundamental Concepts of Metaphysics: World, Finitude, Solitude*, is a translation of the German *Einsamkeit*, which is rendered as "loneliness" in the English translation of Sloterdijk. Sloterdijk notes that the original title of the lecture course used the word *Vereinzelung* ("isolation") and not *Einsamkeit*. Peter Sloterdijk, *Bubbles: Spheres Volume I: Microspherology*, trans. Wieland Hoban (Los Angeles: Semiotext(e), 2011), 649n44.

53. Ibid., 13.
54. Jean-Luc Nancy, *Being Singular Plural*, trans. Robert D. Richardson and Anne E. O'Byrne (Stanford: Stanford University Press, 2000), 3.
55. Such an analysis of Milton's ecological or proto-ecological perspective is given in Ken Hiltner, *Milton and Ecology* (Cambridge: Cambridge University Press, 2003).
56. Many examples of such radical approaches to ecocriticism can be found in Serenella Iovino and Serpil Oppermann, ed., *Material Ecocriticism* (Bloomington: Indiana University Press, 2014). For a sense of how such ecocritical thought engages with the elements, see Jeffrey Jerome Cohen and Lowell Duckert, ed., *Elemental Ecocriticism: Thinking with Earth, Air, Water, and Fire* (Minneapolis: University of Minnesota Press, 2015). Timothy Morton is a contributor in both of those volumes, indicating the overlap between coexistentialism and ecocriticism.
57. Aaron M. Moe, *Zoopoetics: Animals and the Making of Poetry* (Lanham: Lexington Books, 2014).
58. There are many points of convergence between environmental humanities and posthumanities, although the former is not as committed to critiques of humanism and the latter is not as committed to environmental concerns. Both terms are rather equivocal, ambiguous, and contested. This book is allied with both theoretical orientations, although as a work of existentialism, its alliance with any "school of thought," "theory," or "movement" always includes resistance and rebellion, and thus should never be confused with adherence or acceptance. See chapter 7 ("After Humanism") for more.
59. Alfred Kentigern Slewers, ed., *Re-Imagining Nature: Environmental Humanities and Ecosemiotics* (Lanham: Bucknell University Press, 2014). For an account of the evolutionary development of aesthetic and semiotic capacities across human and non-human modes of existence, see Katya Mandoki's analysis of aesthesis and semiosis in three spheres of existence, *Orbis primus* (matter), *Orbis secundus* (life), and *Orbis tertius* (humans). Katya Mandoki, *The Indispensable Excess of the Aesthetic: An Evolution of Sensibility in Nature* (Lanham: Lexington Books, 2015).
60. John D. Caputo, *How to Read Kierkegaard* (New York: W. W. Norton & Company, 2007), 83.
61. Hans Jonas, *The Phenomenon of Life: Toward a Philosophical Biology* (New York: Harper and Row, 1966), 233.
62. Peter Sloterdijk, *You Must Change Your Life: On Anthropotechnics*, trans. Wieland Hoban (Malden: Polity Press, 2013), 452.
63. This multifaceted sense of the underground was explored at the biennial conference of the Association for the Study of Literature and Environment, "Notes from Underground: The Depths of Arts, Culture and Justice," held at the University of Idaho in 2015.
64. Jean-Paul Sartre, *Existentialism Is a Humanism*, trans. Carol Macomber (New Haven: Yale University Press, 2007).

TWO

Existentialist Legacies

Whether you relegate it to the status of an outdated movement or a refuted argument, or you respect it as a decisive influence or a watershed historical moment, or you consider it a convenient label that refers to a disparate group of writers, existentialism is in the past. While precursors to existentialism can be found in Shakespeare, Augustine, and Plato, existentialism itself seems to be a phenomenon of the nineteenth and twentieth centuries, associated with works of people like Kierkegaard, Nietzsche, Dostoevsky, Heidegger, Sartre, and de Beauvoir. However, existentialism is not simply behind the times. It remains to come. You can think of it meteorologically. Like the monsoon in India or El Niño in the Americas, it is a weather system that comes and goes. Although its nineteenth-century emergence subsided by the middle of the twentieth century, an existentialist event is reemerging, gathering on the horizon like a looming storm. A storm can recharge rivers and aquifers, and it can nourish life and land, but it can also bring destruction in the occasion of floods, mudslides, tornadoes, and extreme winds. Like a series of storm cells circulating around the planet, under specific conditions existentialism can produce a storm that recharges theory and practice, percolating into freedom and responsibility, but its critical force can also become destructive or distracted, producing reactive attitudes that wallow in the tepid malaise of cynicism, dismissing with a roll of the eyes any attempt at creative or affirmative engagements in the struggles of coexistence.

For better or for worse, existentialism is not over and done. Its legacies remain, like envoys or ambassadors (*legatis*), diplomatic representatives of existence sending messages about the challenges of living and dying together, the challenges of sharing in the struggle to be. The messages are not just words and phrases. Signs of existentialist legacies can be found in feelings, moods, gestures, elements of existence. They are in the air. You can hear them in the ciphers of a breeze whispering in your ear. They are in stone.[1] You can find them in the strata of the Grand Canyon. You can feel them when a sculpture gives rise to a sense that you must transform yourself, an injunction to change your life.[2] A rising breeze, dropping temperature, darkening sky, and lightning flash. Are those signs of a coming storm? It is not always clear what the messages of existentialist legacies signify. It is not obvious how to prepare for whatever event they convey. What is clear is that the messages are relevant to contemporary challenges of coexistence, challenges of being in the swarming storms of ecological emergency.

Existentialism is arguably more relevant than ever before, particularly in light of the continually intensifying struggles of coexistence, the intensifying destruction and volatility distributed across not only human existence but all of the basic components of life on Earth. Unprecedented uncertainty, suffering, and extinction are traversing oceans, soil, rivers, the climate, and the abundance and diversity of species. Such ecological emergency opens up a space for the arrival of existentialist envoys. In other words, ecological emergency provides rich compost for an ecological renewal of existentialism, not to help people cope with the emergency or be okay with it, as if we could learn to not be bothered by emergency conditions, finding a calm center in the unbearably chaotic storm of coexistence. Indeed, the tendency to avoid being bothered is precisely the problem. You have to bother with emergency conditions if you intend to let something new emerge.

The ecological emergency is not only to be found in species extinction or climate change but also in the apathy and denial that characterize so many reactions to emergency conditions. If the emergency were simply a matter of environmental degradation, then something could be done about it. Emergency conditions could facilitate the emergence of new events, and not necessarily grandiose events like political revolutions. Smaller-scale events count too, like starting a community garden, investing in solar energy research and development, or learning how to grieve and mourn for past, present, and future generations of life. By ignoring the finitude and uncertainty of ecological emergency, the controlling and calculative human eclipses anything different, anything new, anything that could emerge. The most severe emergency is the failure to let this emergency really be an emergency. For Heidegger, this is the emergency of a lack of emergency, "the distress of lack of distress" (*die Not der Notlosigkeit*).[3] The optimism that everything will somehow work out for humans or at least for nature, the pragmatism of policy changes and technological innovations, and the cynicism of critiques against hegemonic systems of capitalism, technoscience, and neocolonialism, those are all shock absorbers preventing the emergency from emerging. I am not arguing against hopeful attitudes, practical actions, and critical thinking. They have their place, but without immersion in emergency, nothing new can happen, nothing different can emerge. Whatever place that hope, action, and critique have is foreclosed by a lack of concern, a lack of *die Not*—need, distress, urgency, emergency. Ecological existentialism is a provocation to find oneself in the storm, affirming planetary coexistence in and as emergency.

Most of this book focuses on the works of contemporary thinkers who are contributing to such an ecological existentialism or, in other words, coexistentialism. I do not attempt to give an ecological reading of the thinkers typically considered part of the existentialist movement. My concern is not to retrieve or reconstruct the ecological elements of those thinkers, which is not to say that there is not value in doing so. My concern in this book is not with existentialism but with its renewal in the ecological thought of the twenty-first century. My concern is coexistentialism. Nonetheless, one cannot engage with ecological existentialism without considering the ideas or energies of existentialism that are undergoing renewal. If you cannot discern existentialist legacies, you cannot discern the difference between the existentialist movement and its ecological renewal, and you cannot discern the difference between ecological existentialism and other ecological perspectives, such as eco-phenomenology, ecofeminism, or ecological hermeneutics. What makes ecological existentialism an *existentialist* movement? What legacies left behind by existentialism are undergoing an ecological renewal?

This chapter responds to those questions by considering five legacies of existentialism: style, experience, resistance, postsecularism, and absurdity. Those are by no means exhaustive, and they are not meant to be definitive, as if it is desirable or even possible to reduce existentialism to an exhaustive or definitive list of traits relevant to contemporary situations. They are more like directions on a map, indicating a way to existentialism. There are surely other ways to arrive there, and the territory is different from the map, so these directions should be taken somewhat lightly, *cum grano salis*. This should not be taken as a theory of existentialism. Along those lines, it is worth heeding the first words that Hannah Arendt spoke during a 1968 seminar: "No theories, forget all theories."[4] If what follows contributes to a theory of existentialism, it is an accidental theory, a theory without theory, a forgetful theory, which has misplaced or displaced the boundary between the theoretical and nontheoretical.

STYLE

Your style indicates your stance, your standing. Your style signifies something about you. It is not just there for survival value. It is more about sexual selection than natural selection, more about exorbitant differentiation and attraction than health and fitness.[5] It shows the unique attunement of your vibrant materiality. In other words, your style is material and semiotic, just as the resonance between the words "style" and "stance" involves a shared sound ("st-") as well as some shared aspects of their meaning, both signifying position, place, and power. For example, your hairstyle has a particular shape, feel, and look, which taken together signify some kind of cultural participation, indicating that you belong to or resist belonging to a particular group. Consider the religious belonging indicated by the haircut known as the tonsure, which involves shaving the top of one's head, while relatively short hair remains on the sides and back of the head. Such a hairstyle could indicate a later stage of male pattern baldness, but it could also be a sign of participation in a monastic community: the tonsure is worn by monks in some sects of Christianity, Buddhism, and Hinduism. A bright green mohawk might indicate resistance to cultural norms, experimentation with self-expression, participation in punk or emo subcultures, or all of the above, not to mention the dubious association of the haircut with a North American indigenous people, the Mohawk Nation.

Your haircut, shirt, shoes, and posture say something about you, the real you, whoever you are. Even if you are dressing up in a costume that completely dissimulates who you really are, your choice of that particular costume and the way you wear it are nonetheless indicative of the person you are. Your style does not say everything about you, but it is not separate from you either. On one hand, you can change clothes and still be the same person, and on the other hand, your changing outfits manifest the dynamic flow of your identity: you cannot put on the same dress twice, as it is a different dress, and you are a different person. Your style does not exhaust the depth of your being, but if you took away all of your style, you would not exist. You may shave off your hair and take off your clothes, but that is still a style.

Philosophical and literary movements are like hair, shoes, belts, and body spray. They assemble into different styles. This does not mean that existentialists all have the same haircuts. For instance, Nietzsche's moustache would prevent him from being grouped alongside the clean-cut faces of Kierkegaard, Camus, and Sartre. Existentialists do not all share the same stylistic accoutrements. They do not all smoke cigarettes, though one might say that they are each committed in various ways to a dangerous lifestyle. This is famously articulated by Nietzsche, who exclaims that "the secret for harvesting from existence the greatest fruitfulness and the greatest enjoyment is—to *live dangerously*! Build your cities on the slopes of Vesuvius! Send your ships into uncharted seas! Live at war with your peers and yourselves!"[6]

One might say that a philosophy or philosophers do not have style. Indeed, many philosophers do indeed appear to lack style. They are buttoned up and boring. That does not mean that they do not have a style. Rather, it means that their style is bland, average, and inconsistent. Having style means having a fashionable style—unique, vibrant, attractive. A bland style among philosophers is often the result of philosophical thinking that suppresses its stylistic elements in favor of some kind of universal neutrality, like trying to talk without an accent. You might sound neutral to yourself or your friends, but there is no speech without accent. Existentialism embraces this. Instead of thinking that a style gets in the way of universality, many existentialists would say that the quest for universality or objectivity gets in the way of the styles of real beings, the existential truths of actual individuals. This is not to say that existentialists reduce philosophy to style, as if it is nothing but fashion. That would be a convenient way to dismiss the philosophical seriousness of existentialism, by reducing it to a fad, superficial aestheticism, a trendy fashion with no content, all style and no substance. Different styles do not exhaustively capture those philosophies and philosophers, but they are not separate from them either. For example, it sometimes seems that you can take the same position using two different philosophical styles, like arguing against mind-body dualism using Alfred North Whitehead's process philosophy or Jacques Derrida's deconstruction. However, you cannot step into the same philosophy twice. As your stance changes, your style changes, and as your style changes, so does your stance. Whitehead's writings posit, define, and explain his ideas with systematic precision, giving you something that you can easily hook up to your own thinking. Derrida's writings open into relentlessly intricate analyses, ornate and perhaps superfluous, never positing without questioning the position of anything that is posited and anyone who posits. To be sure, both of these philosophers exhibit densely crafted, complex prose, but they are not arguing the same thing in different ways. They have different styles. Derrida's style exhibits an exorbitant reflexivity, an abysmal anxiety, a vertiginous *mise en abyme* that contrasts sharply with the systematic diligence of Whitehead's style. The thinking, feeling, and acting expressed in a Derridean response to mind-body dualism would be very different from those expressed in a Whiteheadian response. There would probably be more prayers and tears in the former.[7] Taking on the risk of oversimplifying, one could say that Derrida's style is the more existentialist of the two.

Existentialism bears a specific style. It is a plural and diverse style, with various philosophical, literary, cultural, and artistic iterations. That plurality and diversity might seem to preclude a definition of a specific style of existentialism, as if using the singular noun "style" inherently implies a simple or homogeneous totality to the exclusion of complexity or heterogeneity. The danger of homogenization calls for some hesitation before attempting to gather the many styles of existentialism—many existentialisms—into one style of one existentialism. Speaking paradoxically, the style of existentialism could be described as a style that is not a style, at least not a style that is gathered up into a neatly identifiable list of traits. There is not one fashion of existentialism, one ring to rule over all realms of existentialism. The style of existentialism is found in its openness, a style with no preset style: freestyle. Is it a literary style, a way of using words, or is it a way of being in the world? It is literary and literal, a style of speaking and of being. Yet it is also neither, as its openness does not presuppose a word/world dichotomy. It does not presuppose a conclusive measure of any distinction between words and worlds or between meanings and matters. That would be too safe. Its openness presupposes a sense of uncertainty regarding suppositions and foundations. Who wants stable foundations when you can live dangerously on the slopes of Vesuvius?

Existentialist freestyle comes with a feeling for the groundlessness of what many take to be grounds, and indeed, a feeling for the untenability of any simple distinction between groundless and grounded, not to mention simple distinctions between openness and closure. Consider Kierkegaard's use of pseudonyms throughout his writings. Who is writing? Kierkegaard or Johannes Climacus? Both? Neither? There is no way of knowing with any certainty. Scholarly convention dictates that all Kierkegaard references should be listed under the name Kierkegaard and not under Johannes Climacus, Johannes de Silentio, Anti-Climacus, Constantin Constantinus, or the humorous assembler of publications, Hilarius Bookbinder. Kierkegaard's style is not simply autobiographical. More accurately, his style is to suspend any objective certainty about whether his work is autobiographical. The style of existentialism is not putting yourself into your work. It is the style of holding open the problem of whether one is putting oneself into one's writings, which means holding open questions about whether there is any relationship between oneself and one's writings, questions about whether one has writings (Are there writings? And are they things I have?), and questions about whether one is oneself or even is at all. Along with Kierkegaard's pseudonyms (or is he the pseudonyms' Kierkegaard?), this openness is apparent in another philosopher whose writings are *de rigueur* for any engagement with existentialism, Jean-Paul Sartre. Sloterdijk describes this aspect of Sartre's style, noting that "he remained throughout his life faithful to his way of living the groundless freedom. To him, the nothingness of subjectivity was not a downward-plunging abyss, but a spring bubbling upward," and furthermore, he liked it: "In contrast to many thinkers of subjectivity, Sartre felt comfortable in his abyssality; leaning on anything was for him more a compulsory exercise than free-style."[8] That abyssality is apparent in his portrayal of Antoine Roquentin, the beleaguered protagonist of Sartre's *Nausea,* who is clearly an autobiographical representation of Sartre himself. The novel provides Sartre a space of autobiographical freestyle not unlike pseudonyms for Kierkegaard. Even the details of Roquentin's perceptions are related to Sartre's own intense, personal experiences, such as the crabs and lobster claws flashing into Roquentin's perception during his moments of anxiety and despair, which are easily correlated with similar visions that Sartre experienced during psychedelic experiences with mescaline.[9] However, Roquentin is not Sartre. They are two different names, two different people, one literary and one literal, one fictitious and one factual. Who is this character? Who has nausea? Perhaps both Roquentin and Sartre experience it. Yet, at the same time, perhaps neither does, as neither is completely himself. Their existential pathos oozes out of a rift internal to each. Neither Antoine nor Jean-Paul is himself. Each must become himself, but that act of becoming oneself never reaches any finality or complete closure.

In some sense, Roquentin and Sartre coincide with one another, but in another sense, there is neither Roquentin nor Sartre. All that remains is: nausea. What makes nausea so nauseating is that it is unclear whether it matters if there is anybody there to be nauseous. Consider Roquentin's experience of a chestnut tree while sitting in a park.

> The roots of the chestnut tree were sunk in the ground just under my bench. I couldn't remember it was a root any more. The words had vanished and with them the significance of things, their methods of use, and the feeble points of reference which men have traced on their surface. I was sitting, stooping forward, head bowed, alone in front of this black, knotty mass, entirely beastly, which frightened me. Then I had this vision . . . existence had suddenly unveiled itself. It had lost the harmless look of an abstract category: it was the very paste of things, this root was kneaded into existence. Or rather the root, the park gates, the bench, the sparse grass, all that had vanished: the diversity of things, their individuality, were only an appearance a veneer. This veneer had melted, leaving soft, monstrous masses, all in disorder—naked, in a frightful, obscene nakedness.[10]

He experiences the tree and himself as meaningless, absurd, superfluous, "*In the way*."[11] He considers suicide, which would itself be superfluous, leaving behind a meaningless decomposing body in a meaningless Earth, all of which is *In the way*. That is Sartre's style. He is not simply putting himself into his writing. He does not lean on an identity that could be put into writing. He puts the terrifying abyss that is and is not his own subjectivity into writing, an abyss that opens onto an unbearable intimacy with a tree, the Earth, and all objects. "All these objects . . . how can I explain? They inconvenienced me."[12]

In abysses of self-reflective freestyle, it is not clear what Kierkegaard or Sartre actually thinks, and not because they are trying to hide their personal opinions and leave it up to the reader to decide what to think. Rather, the freestyle is what they think, and it is not clear, as it emerges from ongoing engagements with the exigencies of actual existence. It is a style that never leans on any thought, no matter how true, good, or beautiful. In that sense, the point of an existentialist style is like the point at the end of a stinging fly, like the gladfly to which Socrates compares himself in Plato's *Apology* (30e). The point is not to get you out of words and into the world, or out of writing and into your body, as some phenomenologists would suggest.[13] Rather, the point is to wake you up so that you stop accepting the distribution of beings that you have habituated and begin continually questioning yourself, regularly renewing your freedom by tearing yourself away from yourself. Of course, that would entail questioning the authority that enjoins you to continually question yourself. Jean-François Lyotard gives an apt description of this indeterminate, uncertain style of questioning.

> To think is to question everything, including thought, and question, and the process. To question requires that something happen that reason has not yet known. In thinking, one accepts the occurrence for what it is: "not yet" determined. One does not prejudge it, and there is no security. Peregrinations in the desert. One cannot write without bearing witness to the abyss of time in its coming.[14]

This style of abyssal questioning is apparent throughout the writings of those typically affiliated with existentialism. It entails an intensification of one's everyday life, an intensification that is thoroughly invested in one's writing. Perhaps more than any other figure, Nietzsche is most associated with an ongoing intensification of life infused into writing.

Nietzsche does not have a style but many styles, many vectors of self-transformation, an open-ended plurality of stinging styles, which are traced by Derrida's *Spurs*.[15] Like Sartre, Nietzsche does not lean on any one identity or style. Consider Nietzsche's remarks about *"we aeuronauts of the spirit,"* who are always attempting to fly farther than whatever perch accommodated the previous adventurous spirit:

> All those brave birds which fly out into the distance, into the farthest distance—it is certain! Somewhere or other they will be unable to go on and will perch on a mast or a bare cliff-face—and they will even by thankful for this miserable accommodation! But who could venture to infer from that, that there was *not* an immense open space before them, that they had flown as far as one *could* fly! All our great teachers and predecessors have at last come to a stop and it is not with the noblest or most graceful of gestures that weariness comes to a stop: it will be the same with you and me! But what does that matter to you and me! *Other birds will fly farther*! This insight and faith of ours vies with them in flying up and away; it rises above our heads and above our impotence into the heights and from there surveys the distance and sees before it the flocks of birds which, far stronger than we, still strive whither we have striven, and where everything is sea, sea, sea!—And whither then would we go?[16]

With unparalleled bombastic rhetoric, Nietzsche tears himself away from himself, and he tears us—who?—away from ourselves as well. Nietzsche does not tell you how you ought to be or how all humans ought to be, unless it is in oxymoronic sense of suggesting that you ought to be someone who does not accept uninteresting and vapid distinctions about what anything ought to be. "A man as he *ought* to be: that sounds to us as insipid as 'a tree as it ought to be.'"[17] Transform yourself rather than following the norms imposed externally by those who think they know what one ought to do. Freestyle!

EXPERIENCE AND RESISTANCE

Existentialist freestyling implies a relentless commitment to what is disclosed in the dynamics of actual experience. The word "experience" is vague and not unproblematic, but so is "existentialism." Whatever it means here, "experience" does not refer to a subject experiencing an object. It is something prior to any elaboration of a subject-object distinction. To avoid any confusing or negative connotations, a commitment to experience could be rebranded as a commitment to what opens out of existence, to what exposes itself in the vulnerability of being-with, which is also a commitment to resisting anything that dilutes one's experience, anything that forecloses the ecstatic opening of existence. Imagine Dostoevsky's underground man banging his head against the wall of reason, refusing to accept a simple mathematical truth like two plus two is four. What use are mathematics and the very laws of nature if they foreclose on my freedom, however absurd that might be? What use is truth if it is not lived, experienced, encountered? Consider Kierkegaard's emphasis on existential truth and his criticisms of what was understood as truth by the Hegelians of his day or by the institutions of Christendom. The truths of Hegelianism and Christendom are too abstract. Supposedly true for everyone, those truths are not for any actually existing individual. What do they have to do with me and my life? "What I really need is to be clear about *what I am to do*, not what I must know, except in the way knowledge must precede all action . . . the thing is to find a truth which is truth *for me*, to find *the idea for which I am willing to live and die*."[18] This is not unlike Heidegger's emphasis on facing one's own finitude and mortality in the midst of the inauthenticity that marks the neuter existence of the average person—the anonymous "one," the "They" (*das Man*), the everyday self that does not resolutely choose its existential possibilities for relating to itself, others, and the world.[19]

Affirmation of experience coincide with resistance to abstract, formulaic, and institutionalized structures, like those of philosophy, science, and religion, as well as resistance to the habituated routines of average existence, which descend into meaninglessness. The legacies of existentialism do not only convey the possibility of thinking, feeling, and acting with a free style, they convey the importance of directing that style toward self-enhancement and away from stable grounds that enclose one's freedom and foreclose future transformation. Affirming transformative experience, resisting the temptation to lean on something stable or perch on a temporary foundation, existentialists are envoys of autoplastic aeronautics, spreading the good news about self-transformative practices. Even though many existentialists, with Kierkegaard and Nietzsche as notable examples, were lonely, isolated individuals, living in societies that did not support their freestyling thought, faith, and will, the fact that they disseminated their style in writing attests to their commitment to being with others. As envoys, existentialists point toward the possibility of a transition to coexistentialism, a transition to a community of what Sloterdijk terms "acrobatic existentialism": neither Christendom nor capitalism, it is a community of "travelling people; not a community of saints, but one of acrobats; not paying contributors to an insured society, but members of an organization of those living dangerously."[20]

If living dangerously is something I write about but do not experience, then I am obviously not really living dangerously. Likewise, if I am living dangerously in practices but not in writing, then I still am not really living dangerously. In either case, I would be living safely within a house with neatly partitioned rooms of theory and practice. The house is a historical monument that is old but safe, undergoing regular maintenance and renovation. Theory is a reading and discussion room, and practice is the exercise room. Living dangerously requires becoming uncanny, which could entail getting out of the house, but it could also entail finding a corner to sit in and be bored, getting outside of the usual functions of space in the house. Acrobatic experience resists the comforting confines of convention, whether religious or rational. Regardless of whether the house is made by God or history, living dangerously demands getting outside of the norm. Accordingly, Camus remarks that a historical or rationalist perspective is just as confining, oppressive, and totalizing as the absolute nihilism of an otherworldly God.

> History in its pure form furnishes no value by itself. . . . Systematic violence, or imposed silence, calculation or concerted falsehood become the inevitable rule. Purely historical thought is therefore nihilistic: it wholeheartedly accepts the evil of history and in this way is opposed to rebellion. . . . Cynicism as a political attitude is only logical as a function of absolutist thought; in other words, absolute nihilism on the one hand, absolute rationalism on the other. As for the consequences, there is no difference between the two attitudes. From the moment that they are accepted, the earth becomes a desert.[21]

Colonized by the house of a transcendent God or the house of immanent history, not only is human life devalued. The desertification internal to human life accompanies the spreading desertification of Earth. Another warning of personal and planetary desertification comes from Nietzsche. *"Deserts grow: woe to him who harbours deserts!"*[22]

Resisting desertification, you must resist authority and become an author of yourself, to authorize a sense of the world that is true for you. That kind of freestyling is difficult. Every moment provides temptations to follow conventions, temptations to perch or lean on something. Along with resistance to foundationalisms, maintaining such openness in everyday experiences requires intense and ongoing training in the development of your acrobatic capacities—abilities to experience things (including yourself) differently, opening up to strange horizons, becoming otherwise than whatever is disclosed through habitual or conventional modes of experience.[23] Furthermore, the style of existentialism is not only to maintain this openness throughout one's everyday life but also throughout one's writings. This openness should be distributed across one's literary and literal *corpus*. The writing styles of existentialists are not uniform, but they share an openness that resists foundations and affirms ongoing engagements in the exigencies of actual existence.

Kierkegaard's pseudonyms, Nietzsche's aphorisms, Heidegger's poetic thinking, and the philosophical novels of Dostoevsky, Sartre, and Camus all perform this experimental freestyle, writing to open possibilities and resist the foreclosure of different ways of being. However, pseudonyms or poetic sayings can themselves become foundational perches or walls on which one lands or leans. There are no guarantees on the existential efficacy of any literary devices. You have to maintain vigilance in order to resist closure. You have to get scrappy! I mean both senses of scrappy, something recalcitrant and pugnacious as well as something fragmented into scraps. Kierkegaard's (or Johannes de Silentio's) sense of himself as an *Extra-Scriver*, a "free-lancer" or "supplementary clerk" who does not write for a system (especially not *the* system) but writes for the joy of it. Jot notes on scraps, writing pieces, accounting for whatever falls through the cracks in the ground of the system.[24] This freelance style is also expressed by Donna Haraway, a key figure in the ecological existentialism of Deborah Bird Rose. Playful, tricky, and vitally serious, Haraway says that her writing finds "ways of blocking the closure of a sentence, or of a whole piece, so that it becomes hard to fix its meanings. I like that, and I am committed politically and epistemologically to stylistic work that makes it relatively harder to fix the bottom line."[25] Instead of writing of or for any totalizing system, she repurposes elements of the system to become uncanny and cultivate transformative possibilities. "I tell declarative stories trippingly."[26] Haraway follows what the science fiction author Ursula Le Guin calls a "carrier bag" approach to fiction, which repurposes technoscientific developments, performing a "strange realism" that transforms them into a "cultural carrier bag rather than weapons of domination," a carrier bag not with missiles of destruction but missiles carrying messages of a "strange realism" that imagines different ways of caring for one another, carrying one another in "this womb of things to be and tomb of things that were."[27] However, again, there are no guarantees. Messengers giving and receiving messages can also become a feeding frenzy for meaningless mass consumption. It could be like the situation in Kafka's "Couriers":

> They were offered the choice between becoming kings or couriers. The way children would, they all wanted to be couriers. Therefore there are only couriers who hurry about the world, shouting to each other—since there are no kings—messages that have become meaningless. They would like to put an end to this miserable life of theirs but they dare not because of their oaths of service.[28]

What to do about a cacophony of meaningless messages is a serious problem. Putting kings back into power is not an option though. That is exactly the kind of enclosing power that must be resisted in order to affirm experience and attend to the ecstatic opening of existence. There is one king in particular to which existentialism has been particularly hostile: the most sovereign of all sovereigns, God.

SACRED, SECULAR, ABSURD

Kierkegaard's attacks on Christendom, Nietzsche's proclamation of God's death, the loud and proud atheism of Sartre, de Beauvoir, and Camus, and Heidegger's neopagan spirituality of earth, sky, divinities, and mortals are all examples of the religious dimension of existentialist resistance against conventions and anything that confines the openness of human existence. As Camus noted above, secular rationalism is a continuation of the nihilism of transcendental monotheism by other means, another way of controlling and calculating things, reducing the world to nothing but an inevitable progression, God's plan on one hand, history on the other, both hands turning Earth into a desert. Even though ancient religious modes of existence and modern secularism both seem to head toward an absolutism or foundationalism, existentialists do not always reject both equally. Some existentialists lean more toward the secular and others toward the sacred. Camus himself is much more invested in concepts of history explicated by secular reason than by the demands of faith revealed in a religious tradition. Sartre has no concern for faith in God, whereas Kierkegaard's existential truth is absolutely related to God. Nietzsche's entrancement with Dionysus and Heidegger's attention to poetic sayings of the holy are profoundly religious, yet each of those thinkers is also very critical of institutional forms of religion and traditional conceptions of God.

Whether primarily oriented toward reason or revelation, the sacred or the secular, all existentialists share a refusal of the God of Christendom, the transcendent monotheistic control center. Existentialists who believe in God do not believe in that God, not God as a superbeing who shields humans from the doubt and meaninglessness of existence. They leave behind that God for what the existentialist theologian Paul Tillich calls a "God beyond God" or "God above God."[29] The point is to affirm the event or lived experience of religion while maintaining a critical distance from the metaphysical and institutional sclerosis of religion. The metaphysics of a superhero being and the politics of institutionalizing the power to lord oneself over others both short-circuit the ecstasy of experience, enclosing it within a realm controlled by some theologico-political sovereign. One problem here is that the attempt to replace traditional theologico-political power has been more difficult than some initially estimated. While modernity supplants revelation with reason, secluding the former into private affairs and giving reason free rein in the public sphere, reason thus becomes the new face of foundationalism. Whether rationality is defined in terms of historical development, technoscience, or the free market, it is another perch, another wall to lean on, another way to outsource responsibility for finding truth that is true for you. Accordingly, existentialists struggle to resist the solace of reason and revelation. They are not thereby seeking a dialectical synthesis of the two. Their style can be considered postsecular, as it emerges in the wake of secularism, but it is not postsecular in the sense of a synthesis of sacred and secular perspectives into some comprehensive whole.[30]

Steering clear of the perches offered by the sacred, the secular, and the dialectical integration of the two, existentialists fly out to sea, to the absurd, affirming nonrational elements of experience, altered states of consciousness and nonordinary awareness. Nietzsche has his Dionysian enthusiasm, Kierkegaard has his leap of faith, and Sartre has horror. "How serpentine is this feeling of existence—I unwind it, slowly. . . . If I could keep myself from thinking! I try, and success: my head seems to fill with smoke . . . and then it starts again. . . . At this very moment—it's frightful—if I exist, it is because I am horrified at existing."[31] I am horrified, therefore I am, if I am. Works of horror philosophy are indicative of a radically affirmative gesture, not fear in the face of horror but existential openness to dreadful strangeness, like the openness cultivated by phenomenological attention, as exemplified in Dylan Trigg's phenomenology of horror, or the openness cultivated by mystical contemplation, as exemplified in Eugene Thacker's *Horror of Philosophy* trilogy.[32] Horrified and horrifying, philosophy involves practices of making and unmaking meaning outside the boundaries of reason, drawing from paradox, uncanniness, uncertainty, hallucination, ambiguity, impossibility, *aporia*. Leaps of faith and shocks of horror are two of the multifarious ways of engaging in nonrational modes of participation in the ecstatic openness of existence.

The absurd is not the negation of the rational, and it is not the synthesis of the rational and irrational. It subtends any distribution of the rational and irrational. It is nonrational in the sense that it is beneath the board game where the rational and irrational make their moves. The absurd might be more effectively described as that which is out of tune and unheard of (*absurdus*). The dissonance of the absurd is the anxiety of ecstatic openness. The detuned tune of attunement, the absurd spreads a strange susurration, a weird whispering that is not easily categorized. Like the voice and laughter of Kafka's Odradek, it disturbs the boundary between speech (*logos*) and silence, sense and nonsense, animate and inanimate.[33] The absurd subsists, rumbling beneath the tectonic shifts of traditional and modern senses of the world. It is a relentless murmuring that disturbs the rational refrains of the Enlightenment and the spiritual songs of Romanticism. It is Sisyphus murmuring under his breath while he walks down the mountain to once again and forever roll his stone up the mountain. His murmuring expresses neither revelation nor reason. Exposed to the horror of his fate, he is not expressing exasperation or depression. He is not bemoaning his place in a meaningless universe. His murmuring is not about anything, not expressing anything, not making sense. The murmurs of Sisyphus are joyful noises of struggle—the struggle for existence, the struggle of the shared worlds of the mountain, the stone, and himself. One must imagine Sisyphus murmuring happily.[34]

Perhaps the most joyful noises reverberating far beyond the halls of reason and revelation can be heard in Nietzsche's Dionysian musicality. For Nietzsche, the subordination of music to the *logos* is one of many symptoms of the life-negating force of decadent rationalism that has plagued Western civilization since Socrates. For instance, in Plato's *Republic* (397b–398c), the musical art (*mousiche techne*) that involves lyrical melody (in contrast with other arts performed with the assistance of the Muses, e.g., dance, tragedy, comedy, sculpture, lyric poetry, and various other arts and sciences) is defined in terms of three components: speech (*logos*), harmonic mode, and rhythm. That itself is relatively uncontroversial. It was a common understanding of the Ancient Greeks that musical songs are composed of instrumental harmony and rhythm as well as the poetic sounds of lyrical speech. However, the *Republic* (398d) presents a claim that was quite novel to Plato's contemporaries: "the harmonic mode and the rhythm must follow the speech." In the ideal civilization proposed within this dialogue, it is supposed to be best for the citizens if songs are arranged in a hierarchy that subordinates their harmony and rhythm to their verbal element, their *logos* (401d–402a).

This subordination of the rhythm and harmony of a piece of music to the *logos* has dominated the Western understanding of music. In modernity, one can see the subordination of music to the *logos* in Jean-Jacques Rousseau's argument that music arises out of an imitation of language, not out of an experience of sound or hearing.[35] Furthermore, Rousseau argues that the emergence of music that is independent of speech (instrumental music) is a sign of the degeneration of music insofar as it is no longer contained within the original unity of language at all, in contrast to music with lyrics, where language can still take center stage. Instrumental music is thus doubly removed from originality. If one were to apply this hierarchy to music in light of contemporary developments, one could speculate as to whether Rousseau would approve of the results. Would Rousseau prefer a rap song like The Sugar Hill Gang's "Rapper's Delight" to a classical piece without lyrics like Beethoven's "Moonlight Sonata"?

Aiming to overcome the rationalism of Western music, Nietzsche sought to make way for a life-affirming sense of music, wherein music is not subordinated to speech, but rather, speech emerges from the rumbling forces of music, forces that cannot be contained in reason or religion, Dionysian forces. This argument can be found in his early work, such as *The Birth of Tragedy*, where he claims that the lyrics of Attic tragedy are "sparks of imagery" that pour forth from the musical enchantment inspired by Dionysus, and it extends through his later works, like *Ecce Homo*, where he reflects on the "tremendous hope" that speaks from his overcoming of Western music, a hope for "a dionysian future of music," a hope for a music that springs not from rational language, but from the entrancing, maddening music of Dionysian ecstasy.[36] Moreover, the Dionysian is not necessarily loud or raucous. These rumblings of intense transformation can come in whispers, as you can hear in Nietzsche's Zarathustra: "Then again something said to me as in a whisper: 'It is the stillest words which bring the storm. Thoughts that come on doves' feet guide the world.'"[37] Still words, thoughts on the feet of doves, and perhaps even a strange spool of thread can guide the world. It is Odradek's general muteness punctuated by the occasional rustlings of speech and laughter that brings the storm.

ECOLOGICAL TRANSITIONS

As a movement that ended in the twentieth century, existentialism continues to murmur its messages. The messages do not have meanings, but they are not simply meaningless either. They are notes, some of them from underground, many of them scribbled and scrappy, conveying a sense that I can find a truth that is true for me. The messages do not explain any authoritative position or official stance, but they do show a specific way of being. They show a style, a way of standing without leaning, a way of overcoming your perch. They perform freestyle, experimenting with an open-ended lifestyle oriented toward ongoing self-transformation, which involves attentive adherence to the exigencies of experience, resistance to the conformism and foundationalism that enclose experience or suppress its finitude and mortality, and a postsecular sensibility that rattles the houses of reason and religion with the rumblings of the absurd.

The emphasis on self-transformation and individual freedom in existentialism distinguishes it from coexistentialism, although the stone of Sisyphus, the chestnut tree of Antoine Roquentin, and the wooden liveliness of Odradek are all indications that the "co-"—the withness, the ecological mesh—of coexistentialism is variously intimated and inscribed throughout existentialist writings. What makes coexistentialism different is a matter of degree. Today, new generations inheriting the legacies of existentialism are attempting to fly farther, to get more involved in the struggles of existence on all scales, from a bacterium to a global climate, from a neutrino to the 13.8 billion-year evolution of the cosmos, from suicide to mass extinction.

Subsequent chapters indicate how the style, experience, resistance, postsecularity, and absurdity of existentialism are taken up in various ways by coexistentialists like Deborah Bird Rose, Timothy Morton, and many other contemporary philosophers who do not necessarily call themselves coexistentialists or ecological existentialists but exhibit shared commitments and characteristics with those who do, such as Peter Sloterdijk, Donna Haraway, and Jean-Luc Nancy. There is no official list of who is in and who is out. Furthermore, "in" and "out" are not in a mutually exclusive relationship. Just like the label "existentialism," the people who are supposedly "in" the group called "coexistentialist" might declare that they are out or that they were never in to begin with. Perhaps maintaining exteriority to coexistentialism is the most effective way of inhabiting it. This is celebrated by existentialism, but it is not unique to that movement. The paradoxical complication of being out and being in is an everyday occurrence, as Morton describes:

> Consider being in a doorway: are you inside or outside the room? Consider the status of a poem's title: is it the beginning of the poem or outside of it? Consider a frame: is it where the picture stops, or still part of the picture? Consider a first person narrative. Is the narrator who is telling the story identical with the narrator about whom the story is being told?[38]

Paradoxically, truths are *dialetheias* ("double truths"). In the style of existentialism, being *out is in*. In other words, coexistentialism is out and proud, queering the boundaries marking the struggles for coexistence.[39] The way to be in the group is exactly this: to open up to the outside, to always fly farther. Coexistentialism is an unheard-of fugue, interweaving and exchanging parts with movements outside of existentialism, like phenomenology, hermeneutics, deconstruction, feminism, environmental humanities, ecocriticism, new materialism, object-oriented ontology, queer ecologies, and more, always more. It is the centrifugal movement of existentialism that facilitated its metamorphosis into coexistentialism.

With the complication of who is in or out of coexistentialism, there is no definitive list of similarities and differences between existentialism and coexistentialism. One might sum up a key difference in the ecological focus of the latter. The ecological volatility and devastation of the twenty-first century is of a vastly different scale than what was occurring in the nineteenth and twentieth centuries, with global warming and the current mass extinction of species as two glaring examples of massive problems the specificities of which are unaccounted for by existentialism or by any prior modes of thought. That clearly distinguishes the context in which coexistentialism is emerging, but again, there is no definitive list of similarities and differences between existentialism and its ecological renewal as coexistentialism. To some extent, even the ecological devastation of all life on Earth was addressed by existentialists, who recognized nihilistic tendencies reducing lives, species, and communities to nothing. They warned of the desertification of Earth under way in their time. Nietzsche's words come from the twenty-first century as much as the nineteenth century: "*Deserts grow: woe to him who harbours deserts!*"[40] A difference might be that the nihilism Nietzsche saw as "this uncanniest of all guests" who "stands at the door" of Europe eventually went inside Europe's house and all of the other houses of Earth, spreading everywhere "the *hyperbolic naiveté* of man: positing himself as the meaning and measure of the value of things."[41] The nihilism of this human superiority complex has now entered the planetary house, ravaging the planetary *oikos* through economic globalization and ecological destruction.

The large-scale and collective dynamics of planetary coexistence call for a coexistentialism that renews existentialist legacies while challenging some of its typical characteristics. In contrast to the existentialist focus on individual subjectivity and personal freedom, coexistentialism focuses more on intimacy with others, including human others along lines of gender, sexuality, race, class, ethnicity, religion, age, ability, and more, always more. More than human others, coexistentialism is profoundly concerned with intimate connections that humans share with nonhumans during an era of unprecedented and uncertain ecological connectivity. As a radical ecological movement and philosophical school of thought, coexistentialism reconstructs existentialist affirmations of the exigencies of experience by situating them in the contemporary context of ecological emergency, wherein actually existing humans finds themselves intimately intertwined in complex relations with multifarious actors, animate and inanimate, organic and technological, local and global. Intimacy is not meant to sound like a good thing. At a time of such devastation, it can be horrifying. Indeed, the immensity to which this intimacy opens human existence is unbearable. We are called to carry the impossible, to hold the unbearable, to affirm ecological emergency and do it with style.

NOTES

1. For an ecocritical reading of the challenging agency of stone, see Jeffrey Jerome Cohen, *Stone: An Ecology of the Inhuman* (Minneapolis: University of Minnesota Press, 2015).

2. "You must change your life." Sloterdijk uses this line from Rainer Maria Rilke's poem "Archaic Torso of Apollo" as the title for his book on postsecular training regimes. In the poem, the life-changing command issues from the poet's experience of being seen, exposed to stone sculpted into the torso of a god—Apollo. Sloterdijk, *You Must Change Your Life*, 21.

3. Martin Heidegger, *Contributions to Philosophy (From Enowning)*, trans. Parvis Emad and Kenneth Maly (Bloomington: Indiana University Press, 1999), 75, 166, 168. In this translation, *Not* is rendered as "distress." It could also be translated as "urgency" or "emergency," hence the title of this commentary by Richard Polt, *The Emergency of Being: On Heidegger's* Contributions to Philosophy (Ithaca: Cornell University Press, 2006). On the translation of *die Not der Notlosigkeit*, see Richard Polt, "The Event of Enthinking the Event," in *Companion to Heidegger's* Contributions to Philosophy, ed. Charles E. Scott et al. (Bloomington: Indiana University Press, 2001), 102n37.

4. Hannah Arendt, *The Promise of Politics*, ed. Jerome Kohn (New York: Random House, 2005), xxii.

5. For more on sexual selection in ecological existentialism, see chapter 8 ("Looking Good").

6. Friedrich Nietzsche, *The Gay Science*, trans. Walter Kaufmann (New York: Vintage Books, 1974), 228 (#283).

7. John D. Caputo, *The Prayers and Tears of Jacques Derrida: Religion without Religion* (Bloomington: Indiana University Press, 1997).

8. Peter Sloterdijk, *Philosophical Temperaments: From Plato to Foucault*, trans. Thomas Dunlap (New York: Columbia University Press, 2013), 93.

9. On Sartre's experience with the psychedelic drug mescaline and its relationship with his writings, see Thomas Riedlinger, "Two Classic Trips: Jean-Paul Sartre and Adelle Davis," in *Hallucinogens: A Reader*, ed. Charles Grob (New York: Jeremy P. Tarcher/Putnam, 2002).

10. Jean-Paul Sartre, *Nausea*, trans. Lloyd Alexander (New York: New Directions, 1964), 126–27.

11. Ibid., 128.

12. Ibid., 127.

13. David Abram is a prominent example of one such phenomenologist. David Abram, *The Spell of the Sensuous: Perception and Language in a More-than-Human World* (New York: Vintage Books, 1997).

14. Jean-François Lyotard, *The Inhuman: Reflections on Time*, trans. Geoffrey Bennington and Rachel Bowlby (Stanford: Stanford University Press, 1991), 74.

15. Jacques Derrida, *Spurs: Nietzsche's Styles/Éperons: Les Styles de Nietzsche*, trans. Barbara Harlow (Chicago: University of Chicago Press, 1979).

16. Friedrich Nietzsche, *Daybreak: Thoughts on the Prejudices of Morality*, trans. R. J. Hollingdale (Cambridge: Cambridge University Press, 2000), 228–29 (#575).

17. Friedrich Nietzsche, *The Will to Power*, trans. Walter Kaufmann and R. J. Hollingdale (New York: Vintage Books, 1968), 181 (#332).

18. Kierkegaard, *Papers and Journals*, 32.

19. Martin Heidegger, *Being and Time*, trans. John Macquarrie and Edward Robinson (New York: Harper and Row, 1962), 163–66.

20. Sloterdijk, *You Must Change Your Life*, 62.

21. Albert Camus, *The Rebel: An Essay on Man in Revolt*, trans. Anthony Bower (New York: Vintage Books, 1991), 288–89.

22. Friedrich Nietzsche, *Thus Spoke Zarathustra: A Book for Everyone and No One*, trans. R. J. Hollingdale (New York: Penguin Books, 1969), 319.

23. I elaborate on such training elsewhere. In this book, see chapter 10, "*Askesis*: Shut up and Train!"

24. Søren Kierkegaard, *Fear and Trembling*, trans. Sylvia Walsh (Cambridge: Cambridge University Press, 2006), 5.

25. Donna Haraway, "'There Are Always More Things Going On Than You Thought!' Methodologies as Thinking Technologies: Interview with Donna Haraway," in *Bits of Life: Feminism at the Intersections of Media, Bioscience, and Technology*, ed. Anneke Smelik and Nina Lykke (Seattle: University of Washington Press, 2008), 33–34.

26. Donna Haraway, *The Companion Species Manifesto: Dogs, People, and Significant Otherness* (Chicago: Prickly Paradigm Press, 2003), 64.

27. Ursula K. Le Guin, "The Carrier Bag Theory of Fiction," in *Dancing at the Edge of the World: Thoughts on Words, Women, Places* (New York: Grove Press, 1989), p. 170.

28. Franz Kafka, "Couriers," in *Existentialism*, ed. Robert C. Solomon (New York: Random House, 1974), 167.

29. Paul Tillich, *The Courage To Be*, 2nd ed. (New Haven: Yale University Press, 2000), 186–88.

30. Postsecularism is addressed in the subsequent chapter, "After God, After Nature."

31. Sartre, *Nausea*, 99.

32. Dylan Trigg, *The Thing: A Phenomenology of Horror* (Winchester: Zero Books, 2014). Eugene Thacker, *In the Dust of This Planet: Horror of Philosophy, Vol. 1* (Winchester: Zero Books, 2011); Thacker, *Starry Speculative Corpse: Horror of Philosophy, Vol. 2* (Winchester: Zero Books, 2015); Thacker, *Tentacles Longer Than Night: Horror of Philosophy, Vol. 3* (Winchester: Zero Books, 2015).

33. When asked where it/he lives, Odradek responds, "'No fixed abode,' he says and laughs; but it is only the kind of laughter that has no lungs behind it. It sounds rather like the rustling of fallen leaves. And that is usually the end of the conversation. Even these answers are not always forthcoming; often he stays mute for a long time, as wooden as his appearance." Kafka, "The Cares of a Family Man," 428.

34. "One must imagine Sisyphus happy." Camus, *The Myth of Sisyphus and Other Essays*, 123.

35. Jean-Jacques Rousseau, *On the Origin of Language: Two Essays, Jean-Jacques Rousseau and Johann Gottfried Herder*, trans. John H. Moran and Alexander Gode (Chicago: University of Chicago Press, 1986), 50–71.

36. Friedrich Nietzsche, *The Birth of Tragedy*, trans. Ronald Speirs (Cambridge: Cambridge University Press, 1999), 30; Nietzsche, *Ecce Homo: How to Become What One Is*, trans. R. J. Hollingdale (New York and London: Penguin Books, 1992), 51.

37. Nietzsche, *Thus Spoke Zarathustra*, 168.

38. Timothy Morton, *Realist Magic: Objects, Ontology, Causality* (Ann Arbor: Open Humanities Press, 2013), 29.

39. Coexistentialism can be situated among queer ecologies. Morton discusses the intimacy between ecology and queer theory, which is evident in their intersecting understandings of interconnectedness as a mesh of strange strangers, less of a holistic web and more of an intensification and multiplication of differences. Timothy Morton, "Guest Column: Queer Ecology," *PMLA* 125.2 (2010): 275–77. For a variety of perspectives on queer ecologies, see Catriona Mortimer-Sandilands and Bruce Erickson, eds., *Queer Ecologies: Sex, Nature, Politics, Desire* (Bloomington: Indiana University Press, 2010).

40. Nietzsche, *Thus Spoke Zarathustra*, 319.

41. Nietzsche, *The Will to Power*, 7 (#1), 14 (#12).

THREE

After God, After Nature

Coexist. That word is part of a popular image that shows up on clothing, bumper stickers, refrigerator magnets, hats, posters, and coffee mugs. In that image, each of the letters in "coexist" takes the shape of a symbol. There are different versions, but generally the "c" is a hilal (the crescent moon associated with Islam); the "o" is a peace symbol; the "e" is a combination of astrological symbols (Mars and Venus), or sometimes it is Einstein's equation for mass-energy equivalence ($E=mc^2$); the "x" is the Star of David; the dot of the "i" is a pentagram; the "s" is a yin yang; and the "t" is a crucifix. Coexistence is obviously something that would be of concern to proponents of coexistentialism. However, this sign of coexistence is not what coexistentialism has in mind. First of all, this image is primarily representative of Western traditions. With the hilal, Star of David, and crucifix, the Abrahamic religions are represented. The "e" either represents Western esotericism through astrological symbols or, if it is Einstein's equation, it represents the European scientific tradition. The peace symbol was designed in the 1950s as an antinuclear emblem for the British disarmament movement.[1] The pentagram that forms the dot of the "i" in "coexist"—only the dot, whereas the other symbols shape the entirety of their respective letters—is used in many different contexts, including Wicca, Western esotericism, and Bahá'í. Only one of the seven letters unequivocally represents non-Western traditions, the yin yang "s," the symbol of the ancient Chinese traditions of Confucianism and Daoism. With triumphal finality, emphasizing the Western cultural context for coexistence, that penultimate letter is followed by the cruciform "t," the Christian letter is the ultimate letter.

Coexistence is a real problem facing the contrasting and conflicting faith communities inhabiting the planet together, but the Western ethnocentrism inscribed in the popular "coexist" image is more symptomatic of the problem than it is ameliorative. Furthermore, the problem of coexistence is not only about the coexistence of religious communities, as if it is primarily religions that are responsible for the troubles of cohabitation. Even if the "e" is Einstein's equation, the "coexist" image still focuses too exclusively on religion. Presumably, coexistence involves religious traditions, sciences, arts, politics, and all other areas of human endeavor. Even more, coexistence is not only about many different cultural worlds relating with one another. It is also about worlds relating with Earth, humans relating with nonhumans. The "coexist" image would convey a radically different message if the letters were made up not only of religious symbols but also traced with the signs of DNA, drums, dancers, drills, dingoes, and diamonds.

Coexist! Coexistentialism orients itself toward coexistence, but not necessarily toward the imperative to coexist. A nonprescriptive mode of ethical engagement might more effectively prevent any foreclosure on one's capacities to respond to the demands of actual existence. Framed in terms of an imperative verb, coexistence seems to be something to do, something sought or achieved, treating coexistence like a solution, when in fact it is a problem, an emergency for which there is no solution. It is not something to put on your to-do list. Coexistence is the swarm of beings with which one's existence is enveloped. Coexistence is not something one can attain. It is unattainable in the sense that one is already there. Coexistence is the inevitable condition of anything happening whatsoever. It is also unattainable in the sense that it is unfathomable, a mesh of relationships that is inexhaustible due to its vastness and complexity as well as the irreducible singularity of everything in the mesh. The imperative to coexist makes it seem like the tensions, conflicts, and urgencies of coexistence are problems that can be overcome by finding ways to live in peace and sing in perfect harmony.

Stop fighting and coexist! That is a denigration of the emergency of coexistence, running away from the trouble instead of immersing oneself in emergency to facilitate the emergence of different ways of being together. For Heidegger, this failure to face emergency is the most compelling emergency: *die Not der Notlosigkeit* ("the emergency of a lack of emergency").[2] The most serious danger is the danger of not opening up to the multifarious struggles of coexistence. Gilles Deleuze makes a similar point, adapting Hegel's idea of the developmental stage of morality he calls the "beautiful soul": "The greatest danger is that of lapsing into the representations of a beautiful soul: there are only reconcilable and federative differences, far removed from bloody struggles. The beautiful soul says: we are different, but not opposed."[3] If there is an imperative to coexist, it is not the imperative of the beautiful soul. It is not the imperative to leave emergency and struggle behind, but an imperative to share suffering, to share in the struggles of irreducible differences. Perhaps that is the initial intention behind the popular "coexist" image, to empower religious traditions as they immerse themselves in the ecological emergency. If that is the initial aim, the ethnocentrism and anthropocentrism of the image does not convey the scale of the problem, and the bubbly and cartoonish design fails to indicate an affirmation of emergency. In any case, it is important to address the challenge of facilitating the entrance of religions into engagements with ecological emergency, as is the aim of much work being done in the field of religion and ecology and in the environmental humanities more generally.[4] The ongoing political influence of faith communities in a supposedly secular age is an indication that religions are not going away. Along with political influence, a lot of other things related to religions are also not going away, like daily rituals, funerary practices, nonconceptual and nonrational experiences, community gatherings, celebratory feasts, and storytelling. They may change drastically over time, in the way that the ritual of a morning recitation of prayers morphed into reading the morning paper, which morphed into checking the news online or scrolling through social media feeds. Religious practices, feelings, institutions, beliefs, experiences, and stories are not going away. No matter how troubling they are to those inside of their groups as well as those outside, it would be a mistake, an ecological violence, to enjoin religions to reconcile their differences.

With its postsecular flare for the absurd, coexistentialism inhabits the intersection of religion and ecology, opening that intersection up to an abyss of differences, without foundations, without God, without nature, without the federative harmonies of the beautiful soul. However, since being out is in, the implications of that "without" are not merely privative.[5] The logic of the "without" is what is important here. The point is not to prohibit the use of the words like "God" and "nature," marking off the specific words that you are allowed to say, as if you are in the know, adept at the most cutting-edge theoretical trends available. The point is to kindle a passion for paradox, holding in abeyance any accepted conventions and going to the limit where different possibilities open up. The without is the outside that is also inside, the difference that is also continuity, the out that is also with. It is the frame, which is and is not part of the artwork, or the title of a poem, which is and is not part of the poem.

To approach religion and ecology without God and nature means approaching the limit where religion and ecology are vulnerable to emergency, where different events might emerge and grant new possibilities. Coexistentialists are after that emergency, seeking the emergence of new possibilities. Accordingly, along with saying "without God" or "without nature," a better phrase might be "after God, after nature."[6] The word "after" has two senses here, following Mark C. Taylor's sense of inhabiting a place after God. "On the one hand, to come after is to be subsequent to what previously has been, and on the other hand, to be after is to be in pursuit of what lies ahead. . . . Neither here nor there, neither present nor absent, after is the trace of that which gives and takes being."[7] Being after

God and nature is about being subsequent to God, nature, and all totalizing categories that sunder the world into two clear and distinct categories of transcendence and immanence. The complexities of planetary coexistence have rendered those realms untenable, as indicated by numerous announcements of the deaths of God and nature. Coexistentialism is also after God and nature in the sense of seeking or tracking God and nature, following God and nature as they slip outside of inherited dualisms of natural/cultural, material/divine, physical/spiritual, etc. Being after God and nature in both of those senses, coexistentialism opens up possibilities for an updated form of animism—a postsecular animism for our planetary era, an animism without animism, in other words, an animism "under erasure" (*sous rature*).

SUBSEQUENT

The death of God has been written about extensively since Nietzsche's proclamation that "God is dead."[8] If one considers all of the writings on the death of the incarnation of God as Jesus Christ, then the writings on the death of God multiply immensely. They multiply again if one considers the deaths of other gods, like Osiris, Dionysus, or Quetzalcoatl. To add to this further, some gods and goddesses are not simply dead. They are death, like Yama and Kali. It seems that all God-talk (*theo-logia*) includes talk about death, God, and the death of God. Theology is always already theothanatology, haunted by death, dying, and the dead. In that case, there is something banal about Nietzsche's announcement of God's death, contrary to Nietzsche's own insistence on the unique significance of his message. There are thus at least two ways that one can read Nietzsche's account of the death of God. One can read it as a banal repetition of a typical or archetypal religious image—another dead deity—or, alternatively, one can read it as an announcement harkening the arrival of a new event that has disastrous consequences the likes of which humans have never seen.

Reading Nietzsche's announcement according to its banality, one could say that reports of God's death have been greatly exaggerated by Nietzsche and his acolytes in God's funeral procession. God is dead? That is not remarkable, as God dies all the time. God is always traversing life and death. As Martin Buber puts it, the death of God is really more of an eclipse, part of a dynamic of revealing and concealing that depends upon whether humans have an attitude of intimate relationality with God (an I-You relation) or an attitude of impersonal reification (an I-It relation).[9] In other words, God is still there, buried beneath all of the objectifications whereby humans understand themselves, their world, and the ground of their existence. We can still find God, and we can do it by cultivating I-You relations in the three "spheres in which the world of relation arises," other humans, artworks, and the natural world.[10] If what was announced as death is really an eclipse, then nothing really disappeared. The sun was not unchained from Earth, leaving both to spin indeterminately, with no measure of up or down. That is only an optical illusion, not affecting the reality in itself. "An eclipse of the sun is something that occurs between the sun and our eyes, not in the sun itself."[11] It appears as though human beings no longer see the light of God. The phrase "God is dead" is hyperbole, not the announcement of real extinction. "The eclipse of the light of God is no extinction; even tomorrow that which has stepped in between may give way."[12]

Nietzsche's announcement is thus not the end of the world. That does not mean that it is unimportant. Nietzsche's proclamation is a compelling call for humans to realign themselves with intimate relationships and to stop objectifying things. That is an important contribution, but it is not original or unique. Such calls happen all the time. They are part of an ongoing effort of humans to maintain I-You relations and avoid the ossification, sclerosis, and objectification whereby every You becomes nothing but an It. I would not disagree with an interpretation that emphasizes the banality of the death of God. Insofar as it announced a dead God, it is not new, and likewise, there is nothing new about calling for humans to live vibrant lives full of intense encounters instead of settling into the ruts of habituated experience. However, a reading that emphasizes the banality of Nietzsche does not preclude a reading that emphasizes its staggering uniqueness. To some extent, both readings are accurate. On the one hand, Nietzsche does repeat gestures that have been expressed many times before him in many different cultural and historical contexts, and in that sense his decomposing God reeks of banality, yet on the other hand, Nietzsche's repetition is different, radically different.

If one reads Nietzsche's announcement as harkening a new event, then the death of God is not just an eclipsed reality to which humans can reorient or reawaken themselves. Indeed, if God is lost, what humanity remains such that humans could act to reorient themselves as humans to their lost God? Losing God means losing the ground of humanity whereby humans could act as humans. In other words, not only is there nothing to which humans can reawaken themselves, there is not even a human. The death of God is also the death of the subject, the human agent who has a horizon within which action is possible. As Foucault famously states, "man is an invention of recent date," and if the arrangements of human knowledges crumbled, "man would be erased like a face drawn in the sand at the edge of the sea."[13] One might protest, saying that a human agent still exists. There is still a subject. It is quite obvious: I can still use a subjective pronoun, "I," for instance, so there must be someone there. Think again. Who is saying "I" and "You"? In the absence of the source of all meaning and value, how can pronouns or any other words make any sense? Just because you can say the word "You" and other people can understand it does not imply that it means anything. What is grammar if God is gone? If you still believe in grammar, you are not encountering the event to which Nietzsche is calling attention. "I'm afraid we're not rid of God because we still believe in grammar."[14]

Along with God, all spheres of relation collapse. Unchained, the sun and Earth are both adrift in nothing. With the transcendent realm undone, so too is the immanent world. "*Along with the true world, we have also done away with the apparent!*"[15] As Sloterdijk puts it, Nietzsche's message announces a shock or traumatic event that "undermined the psychocosmic immune system of Old Europe," leaving God, human, and the world in ruins, breaking the metaphysical container that held Earth and sky together in intimate communion: "the desert is growing, the point of orientation is lost," Earth's rotation is that of "a fatal centrifuge hurling us into an eternal cold, and the absence of the sky's final canopy is assumed to bring about the de-securing of life."[16] The loss of God and the desertification of Earth converge in a catastrophic event of nihilism. The death of God accompanies the death of nature, a death announced by many environmentalists, as in the climate activist Bill McKibben's *The End of Nature* and in the environmental historian Carolyn Merchant's classic ecofeminist text *The Death of Nature.*[17] Of course, there are still flora, fauna, elements, and minerals on Earth. Perhaps Buber would say that the death of nature is actually an eclipse of nature, with technoscientific control obstructing human encounters with nature. It might seem that, while natural beings are undergoing real problems, including real extinctions, nature itself is not dead, at least not all of it, not yet. However, although eagles, ospreys, chestnut trees, spiders, and blue-green algae exist, all of those things have lost their background or horizon, becoming displaced from a common reference point or goal. The ecological mesh of organisms and environmental conditions persists, but it does so without any background, without a unifying field, without a ground, whether transcendent or immanent. This is what Morton refers to as "ecology without nature."[18]

The loss of the transcendent realm of truth occupied by God is also a loss of the loss of the immanent world grounded in God, the world of matter, nature, and sense. The point is not that the ground has opened up into a groundless ground, as if the problem was simply that the ground had been conceived of as simple and not complex, monotheistic instead of panentheistic, closed and not open, a noun and not a verb, a static substance and not a dynamic process. Who or what—or, more undecidably, "whot"—remains?[19] What comes after God, after nature? It is not a mere nothing that remains. The ground, that which is grounded, and the act of grounding are all ground into dust. Consider Michael Marder's remarks. "Where there once was a world, there is now dust, which is a world in its own right."[20] Dust is not nothing. It is a swarm of singularities, a crowd of liminal actors. "Dust crosses the boundaries between the living and the dead, plant and animal matter, the earth and the sky, the inside and the outside, you and the world ("for dust thou art, and unto dust shalt thou return"). What does it smuggle in each of these crossings?"[21] Covered in dust, as dust, the background/foreground distinction is indiscernible. Nature is not an eclipsed background. It is not a dynamic and spontaneous process hidden by the objectifying tendencies of scientific rationality or the language games of nouns and verbs.

There is no nature, or if there is nature or one wants to continue using the word "nature," nature must be understood as profoundly weird. The object-oriented philosopher Graham Harman summarizes this point wonderfully.

> Nature is not natural and can never be naturalized, even when human beings are far from the scene. *Nature is unnatural*, if the word "natural" is meant to describe the status of extant slabs of inert matter. The life of gravel and sandpaper is every bit as troubled by inner ambiguities as human existence ever was.[22]

That is a nature without nature, a nature ground into the dust of the myriad things, each of which harbors an inexhaustible depth, like the grain of sand in which William Blake sees a whole world.[23] Gravel, sand, and sandpaper are as lively, bizarre, and irreducibly singular as mushrooms, salmon, potatoes, cats, and humans. The singularity of the human experience is not exclusively human, as Nancy reminds us.

> "Nature" is also "strange," and we exist there; we exist in it in the mode of a constantly renewed singularity, whether the singularity of the diversity and disparity of our senses or that of the disconcerting profusion of nature's species or its various metamorphoses into "technology." Then again, we say "strange," "odd," "curious," "disconcerting" *about* all of being.[24]

The emphasized "*about*" indicates that saying "strange" about all of being means attributing strangeness *to* being, and it means saying *around* being, circumnavigating being without ever gaining access to it. We say strange, then, *about* all of human existence, all of nature, and indeed, all of being. It is all so strange that when we speak of it, we speak around it, as our speech is warped and twisted by the impenetrable, unassimilable, irreducible singularity of the things of which we speak.

While the death of living beings, extinction of species, degradation of ecosystems, and destabilization of the climate are real problems, unbearably real, the loss of nature is good. The death of God and the death of nature are emancipatory events. With them gone, it is possible to engage in the perplexing complexity of events that weave together knots of nature and culture, ecology and religion, and science and politics. Latour does not mince words on this point:

> Thank God, nature is going to die. Yes, the great Pan is dead. After the death of God and the death of man, nature, too, had to give up the ghost. It was time: we were about to be unable to engage in politics any more at all.[25]

With God and nature out of the way, it is possible to practice ecology, studying the struggles for coexistence without forcing things into the boxes of transcendence and immanence. Those boxes do not only take the shape of classical conceptions of God and nature. They take many shapes. When someone reduces a being to its material substrate, the ecology of coexistence is short-circuited by nature. The denaturalization offered by much poststructuralist theory is no better, as it reduces a being to its place within arrangements of discursive practices and power regimes. Society, the new God, becomes the creator. To distinguish it from the connotations that "denaturalization" has with social construction, perhaps a better term for the coexistentialist perspective would be "post-natural."

SEEKING

After God, after nature, is coexistentialism a post-natural post-theism? In some sense, the answer is yes. Coexistentialism is closely connected *with post*al thinking, com-post. Coexistentialism is situated subsequent to the nihilistic catastrophe of modernity, marked by the destruction of the psychocosmic immune system holding together God, nature, and humans, and the destruction of countless real beings—lives, species, persons, places—amid massively distributed endangerment and extinction. Yet, a postal position is in some ways too programmatic and systematic. If a coexistentialist has a position working for a postal system, it would likely be in the Dead Letter Office, which is where all of the undeliverable messages end up, the impossible messages, with addresses and return addresses that are invalid, illegible, or some combination thereof. Scribbled scraps and indecipherable messages are fashionable stylistic accoutrements handed down from generations of existentialists. Coexistentialists are relaying messages of the impossible in a wider, ecological distribution, relaying impossible letters—messages *about* the singularity of all others, all agents in the swarm of coexistence: nature, humans, God, all beings. . . . Some departments in the philosophical postal system are getting a lot of messages delivered, with coherent passages from origins to goals, from the pre through the post, with messages addressing humanism, industrialization, modernism, metaphysics, colonialism, feminism, secularism, and so on. Regardless of whether there are any merits to describing historical or cultural trends and patterns in the linear or dialectical terms of the postal system, one is left wondering about those letters that do not quite make enough sense to get delivered, the dead letters, the spooky and spectral letters of the impossible. Coexistentialists look for those letters, to collect and relay that which could not be properly collected and relayed.

After God, and after nature, coexistentialists are still seeking God and nature, still after God and nature, seeking the religious messages of God without God, seeking the ecological messages of nature without nature. Coexistentialism is after the remains of God and nature, the dust swirling where there was once a world. Whither religion and ecology if all that remains is just dust? Perhaps the easiest and most common reply is denial. It is not all just dust. What remains if not just dust? To what reference point are you orienting your life during this time of psychocosmic immune deficiency? Increasingly, people who feel the inadequacy of the intimacy and meaning provided by religious traditions are turning to spirituality. Many contemporary seekers are finding contemporary spirituality. From an existentialist perspective, it would be far better to realize that what remains is just dust than to be spiritual, at least in the sense with which most self-proclaimed "spiritual" people use the word "spiritual," which is often a shortened form of "spiritual-but-not-religious." It is far better to be "dispirited," as David Webster puts it in the title of his analysis of spirituality, *Dispirited: How Spirituality Makes Us Stupid, Selfish, and Unhappy*. In an existentialist style, he gives an aggressive attack on spirituality. "When someone tells me that they are not really religious, but that they are a very spiritual person, I want to punch their face. Hard."[26]

Webster's overall thesis is correct. Contemporary spirituality does indeed exhibit a tendency to make those who adhere to it stupid, selfish, and unhappy. It is important to clarify what is being said here, as these are obviously incendiary remarks, but not merely so. Webster is careful to say that he is talking about contemporary spirituality in general, a very loosely defined movement that affirms the meaning-making and personal growth associated with the spiritual quest without all of the baggage involved with a commitment to any particular religious tradition. It typically involves finding your own unique path by mixing and matching parts of religious traditions and scientific facts that you feel suit you. There is often a proclivity for the secrets of esoteric traditions and the "new age" optimism of developmental-evolutionary models. There is much contrast and diversity. One can use the word "spiritual" in almost any way one pleases. That individually catered style is a prominent part of contemporary spirituality. In any case, it is not all the same, and it is not all stupid, selfish, and unhappy. Webster acknowledges that some people use the word "spiritual" or "spirituality" in ways that are associated with affirmative, life-enhancing modes of existence. A notable example along those lines is Pierre Hadot, for whom "spiritual exercises" are part of philosophy as a way of life.[27] Aside from relatively rare instances of spirituality like Hadot's spiritual exercises, contemporary spirituality in general does contribute to stupidity, selfishness, and unhappiness.

There is a cafeteria style of pluralism and syncretism that encourages adherents of contemporary spirituality to pick and choose what they want and do not want from each tradition, even picking things that are not compatible, like Hindu meditations on the identity of the self (*Atman*) with the absolute (*Brahman*) and Buddhist meditations on the reality that there is no self (*Anatman*). This apparent inclusivity and openness harbors some stupidity, which is to say, anti-intellectualism. Debates between competing truth claims are not frequent occurrences among spiritual people. Debates are for academics and scholars, who are spiritually immature reductionists deluded by calculations and blinded by reason. Alternatively, debates are for religious people, who are dogmatic, fanatical, or both, debating only to defend their authority. Taking the approach of the beautiful soul, a spiritual person sees no need for debates. There is only harmony and a field of reconcilable differences: if there is a conflict between Hindu and Buddhist truth claims, I can still mix their practices as I see fit, because I found their deeper truth beneath the superficial differences. Spiritual people transcend debates by participating in the perennial truth of mystic unity. The different surfaces of religions can be sloughed off like so much dead skin. In short, spirituality is "faith-lite."[28] Like Kierkegaard's "leap of faith," religion makes heavy demands on practitioners to adjust their thinking, feeling, and acting to attune with what has been revealed according to the tradition. Spirituality does not make such demands. Under the cozy covers of personal freedom and choice, it smuggles stupidity and intellectual laziness. To make matters worse, it also makes people selfish.

Whether they are religious or secular, social and political institutions are forces of community building and organizing as well as disorganization and destruction. Spirituality is about a personally designed inner journey to the true nature of the self, the unity of the universe, and the secret mystery of reality. That privileging of personally designed spirituality orients spiritual people in a way that tends to be less engaged in social and political activities. In short, it makes them selfish. It is not just that spirituality avoids politics grounded in religion in favor of secular politics. Some spiritual people do, such as those involved with the Network of Spiritual Progressives, addressing spiritual values while maintaining the secular prohibition against religion in the public sphere. However, for the most part, contemporary spirituality avoids politics grounded in religion as well as secular politics grounded in reason and representation. On a personally designed journey, the spiritual person does not have to pay attention to debates about truth or participate in the constitution of society. Put simply, contemporary spirituality is selfish, and even if it cares about the cosmos, it is not ecologically engaged. It is otherwordly in the sense that it is disengaged from the troubles of existing together in this world. "Spirituality cedes the world to the worldly."[29]

Ironically, while critiquing consumerist materialism for desiring material possessions instead of spiritual realization, the main kind of community that spirituality tends to support is consumerist community. Spirituality is entirely complicit within capitalist modes of social and political organization. Adherents of contemporary spirituality are buying the same kind of magazines, driving and flying to the same meditation retreats and couples workshops, patronizing similar therapists and life-coaches, using the same websites. Spirituality sells happiness, but that happiness is not authentic, at least no more authentic than the happiness I purchase when I buy a guitar or buy a book criticizing capitalism. Webster follows the existentialist approach to happiness, for which happiness can only be attained by resolutely facing our own finitude and mortality, including the sociopolitical burdens of being finite and mortal together. "If Camus can imagine Sisyphus happy, we too can turn our shoulders to the boulder and get stuck into living."[30] Spirituality is not overwhelmed with fear and trembling in the face of existence. Instead, spirituality enjoins us to ignore the abyss and follow our bliss with easy answers about oneness, immortality, and positive thinking. When happiness comes from within, you no longer have to worry about the countless human and nonhuman others suffering outside of your personal bubble. You do not have to worry about your complicity in the system that is propagating their suffering. Effectively privatized, happiness becomes the name for a rather unjust and miserable way to live, a synonym for destructive delusion.

Although Webster promotes existentialism as a viable alternative to spirituality, his existentialism sounds too humanistic for a coexistentialist intervention at the intersection of religion and ecology after God and nature. To his credit, he distances his position from humanism in the concluding two paragraphs of the book, but not very far. He only distances himself from the British Humanist Association's version of humanism, but he still reduces the religious project to a human invention. Webster claims that, if we want or need rituals, "then we are as capable of inventing them without spirituality and religion as we have been when we invented them as part of religious traditions."[31] Who is this rather capable "we"? It sounds like it uncritically presupposes humans as the inventors, as if the plants, animals, seasons, songs, and dances involved in rituals are so involved solely because of human inventiveness, and not because of any other-than-human power. "Outsourcing mental activities" can indeed move us "further from our ability to choose, from our agency."[32] However, restricting agency entirely to humans moves us further from our ability to care for nonhuman modes of existence. What about the agencies of plants and animals or the agencies distributed across multispecies networks? In addition, and this is particularly problematic for any account of religion or spirituality, Webster promotes "being atheists today," and in doing so, he fails to give any account of the ontological weight of angels, spirits, and God.[33] Maybe angels and deities do not exist, but how do they have the impacts and effects they do? People do not just *believe* in angels and divine revelations. What about people, religious, spiritual, or otherwise, who are actually having *visions* or *encounters* with these things? Are we just going to chalk it up to hallucination? Imagination? If so, what is the ontological significance of a hallucination or of image? Even if they are hallucinations, they are still powerful, and not something to be dismissed as if they are always and merely a facade for anti-intellectualism or death-denial.

Whither religion and ecology if all that remains is just dust? Clearly spirituality is not the answer. Spiritual people retreat into the personally designed bubbles of anti-intellectual, apolitical, consumerist individualism. Suffering from psychocosmic immune deficiency, they disengage from anything in the swirling dust of coexistence, religion, ecology, politics, anything. Webster's existential humanism is better at finding joy in taking up the struggle to exist, but it fails to affirm the nonhuman agencies at work in the struggles of existence. For Webster, "we [humans] are alone together; just people, with no spirits, angels, collective mind-force, gods or dead ancestors," and the space of this aloneness is not meaninglessness but "an absence for us to fill ourselves with a world of value and meaning of our choosing."[34] Webster has not fully felt the shock of existence without God and nature. We are not just people. With the deaths of God and nature, there is no collective "people," no universal human subject. A more ecological account would suggest that the absence left by the death of God, spirits, angels, etc., is to be filled with a world of value and meaning not of human choosing, a human ethics, but of a choosing distributed across all of the planet, a planetary ethics. In that case, meaning-making would be understood as a much more open-ended process. Not located simply in the human, meaning-making is evident in the self-organizing dynamics evident throughout cosmological and biological evolution.[35]

We are not just people. We are and we are not. We humans are entangled in multispecies networks emerging amid the complex, self-organizing dynamics of the cosmos, and our personhood is not that of a homogeneous people but is radically multiple and heterogeneous. Subject positions refract along multifarious lines of difference, such as race, class, gender, sexuality, age, ability, ethnicity, nationality, and so much more, always more. By shifting from anthropocentric to planetary ethics, which entails redistributing agency in ecologies of self-organizing dynamics and diverse subject positions, Webster's existentialism would thus more closely resemble coexistentialism. Such an approach to planetary ethics at the intersection of religion and ecology is proposed by Whitney Bauman. There is no omnipotent God or background nature in Bauman's engagement with the intersection of religion and ecology. Bauman develops a planetary ethic through the articulation of "viable agnostic meaning-making practices."[36]

Instead of the selfish stupidity of spirituality or the powerful knowledge claims of religious and scientific perspectives, Bauman proposes a viable, agnostic theological and epistemological method of opening up meaning-making practices to the edges of certainty and sense in order to facilitate the emergence of caring, loving relationship with Earth's places and their human and nonhuman inhabitants. Agnosticism maintains suspenseful, open-ended engagements with contrasting ways of meaning-making, including ways of making meaning that involve deities and angels as well as ways that involve scientific instruments and experimentation, political participation and representation, artistic expression, gardening, sports, and any other activity that generates meaning, including the values generated in the self-organizing dynamics that pervade the universe, from the most basic processes of matter-energy to the most complex organisms. Perhaps the best way to participate in coexistence after God and nature is to finally admit that we humans do not know the best way to participate in coexistence. We do not know as much as we often pretend to know, and we are not in control, at least not to the exaggerated extent to which humans with power and privilege at their disposal tend to think that they have control.

Bauman's viable agnostic method is a good example of a coexistentialist approach to religion and ecology. However, a coexistentialist might be too immersed in paradox and ambiguity to commit to viability or agnosticism. With a passion for paradox, it is difficult to secure any determinate sense of the boundary between the viable and the inviable. Sustainability, surviving, thriving, and flourishing all sound desirable, but a commitment to the struggles of coexistence entails a commitment to finitude and mortality, a commitment not only to livability or viability but also a commitment to sharing suffering and sharing death. Viability turns its back on the ultimately inviolable unviability of existence, i.e., finitude. I am not saying that anyone should give up the struggle to mitigate climate change or reduce air pollution, but I am saying that turning your back on death is not the way to do it. The "hippie aesthetic of life over death" is not the way to go.[37] Nothing lives. That is not opposed to life. It is its condition. I am already dead. Consider Nietzsche's advice. "Let us beware of saying that death is opposed to life. The living is merely a type of what is dead, and a very rare type."[38] Rose makes a related point: "life and death are mutually interactive . . . and we are thrown back on what existential philosophy has called the Absurd."[39] Oppositions covered in dust become indiscernible: living and nonliving, animate and inanimate, Odradek and a spool of thread.

Along with propagating a false opposition between life and death, to put religions in the service of viability is to cover the singularity of religious perspectives with prescriptive claims, which compromises the agnosticism of the project with the certainty that we must seek viability. Moreover, although Bauman's agnosticism is preferable to Webster's atheism insofar as the former is more open to uncertainty and more hospitable to multiple meaning-making practices, it is not sufficiently ambiguous, weird, or paradoxical for coexistentialism. His agnosticism is too certain about the problems of certainty and knowing and the promise of uncertainty and unknowing. *Gnosis* is not always something to critique or avoid. Consider Morton's notion of ecognosis.[40] It is an interpretive way of knowing without knowing.

> Ecognosis is like knowing, but more like letting be known. It is something like coexisting. It is like becoming accustomed to something strange, yet it is also becoming accustomed to strangeness that doesn't become less strange through acclimation. Ecognosis is like a knowing that knows itself. Knowing in a loop—a weird knowing.[41]

Ecognosis is about realizing one's intimate intertwining with nonhumans and humans through practices of weird attunement, performing the detuned (*absurdus*) tune of ecstatic openness to others.

Affirming the struggles of coexistence is less about maintaining uncertainty about different meaning-making practices and more about attuning with the undecidable who or what of everything in the swirling dust. Overwhelmed by the swarm of coexistence, containing multitudes, I am not sure if I am agnostic. I cannot decide if I know, or if I believe, or if I can decide. Do I believe? I cannot decide, yet I am nonetheless called to believe, to know, to attune. I am after God and after nature, subsequent and seeking, too late and too early, flying farther into the open, beyond the current coordinates that mark the boundary between livable and unlivable, between knowledge and agnosticism, theism and atheism, meaning and meaninglessness, making and unmaking.

Just like life and death, let us beware of saying that meaninglessness is opposed to meaning or certainty is opposed to uncertainty. Perhaps a nonprescriptive liminal method would help open up Bauman's (un)viable (a)gnosticism. Perhaps religion is better defined in terms of practice or training and not in terms of the production of sense or meaning-making.[42] In any case, the point is that a viable agnostic approach covers the swarm of coexistence with too many layers of meaning-making, unjustly assimilating the singularity of what remains after God and nature—just dust. As Marder observes in his writing on dust:

> Dust is just because it restores to beings their own, putting them in their rightful places within the vast expanse of matter. Stated this way, the two senses of *just*—the fact of bare existence and the norm of evenhandedness—overlap. Mereness is justice, to the extent that it keeps each thing within its limits, guarding against the distortions of idealization. Justice is mereness, insofar as it simply sorts things out according

> to the regions of being where they already belong in light of what they are. Injustice, for its part, is the extraneous addition of stratum upon stratum of meaning to the spatiotemporal existence of things.[43]

This is not a statement of a banality or a homogeneous equality of things. It is a statement about their swirling singularity. We are not just "people," and we are not just heterogeneous multispecies events immersed in the meaning-making processes of vibrant matter. Just dust remains: undecidable, without God, without nature, without prescriptions. If you care about coexistence, then you are probably less interested in navigating meaning-making practices in the service of viability and more interested in struggling to be with one another, irreducibly strange, exposed and vulnerable, in the dust of this planet.

ANIMISM WITHOUT ANIMISM

Eugene Thacker uses the image of "the dust of this planet" to convey the singular strangeness of planetary existence. The alterity of planetary dust is an abject horror. Neither the subjective *World* ("world-for-us") nor the objective *Earth* ("world-in-itself"), the planet that remains after the collapse of God, nature, and humans is "the world-without-us."[44] The horror of the dust of this planet is not a problem to which a solution must be sought. It is the emergency in and as which the struggle for ecological coexistence emerges. It is the abyss without which there is no future possibility of a world, for us, in itself, or otherwise. Affirming the dust of this planet is a practice of horror not only in a literary sense or in terms of an emotion. It is a practice of "darkness mysticism," "negative philosophy," or "non-philosophy," a practice of affirming "the paradoxical thought of the unthinkable."[45] It is not a theological mysticism but a "*climatological*" mysticism, "a mysticism of the unhuman" after God and nature, "a kind of mysticism that can only be expressed in the dust of this planet."[46]

The vertiginous multiplicity of the swirling dust of coexistence is not amenable to classical oppositions between subject and object, the one and the many, identity and difference. Accordingly, a problem with relating to the dust of this planet through mysticism would be the tendency for mysticism to become monistic or unitive. Thacker's horror would surely scare off most monists. Anyone seeking to bring the dust of this planet into the light of a holy whole would find themselves falling through an abysmal hole of darkness. Nonetheless, the monistic connotations of mysticism are problematic. Mysticism risks quickly gathering up all differences into an All. The All of darkness mysticism is obviously closer to the legacies of existentialism than the All of optimistic and theistic mysticisms, but it is still an All—a ground or background that provides a context for the dust of this planet.

The tendency to reduce or assimilate the swarm of coexistence into a One or All is a common problem, not unique to mysticism. Close by are other attempts to gather dust into the dustpan of the All, such as panpsychisms, pantheisms, and panentheisms. There is much merit to their efforts to weave soul and divinity together with bodies and the natural world, particularly insofar as those efforts work to overcome the transcendence/immanence dichotomy that persists in so much philosophy and religion. However, the dustpan of the All short-circuits effective companioning with the singularity of other beings, turning them into manifestations of a common ground, different shores on an ocean of oneness. This is not to say that something more pluralistic is called for. It would have to be a specific pluralism, one which does not efface the singularities of the swarm.

Pluralism is equivocal. Sometimes it is only a multicultural or epistemological pluralism that leaves intact that unity of some primary ground like nature, world, matter, or God. Sometimes pluralism declares itself ontological, but that could mean either a pluralism of ideas and beliefs about being, which is another way of restating epistemological pluralism, or it could be a pluralism of modes of being. If it is a pluralism of modes of being, it still does not necessarily affirm the dusty swarm of coexistence. Ontological pluralisms often unjustly inject degrees or layers of interiority, subjectivity, or intrinsic value, thus extraneously adding stratum upon stratum of meaning to what is: just dust. Furthermore, pluralisms often smuggle dualisms of the one and the many, where the many is seen as primary and oneness is an epiphenomenon, abstraction, or aspiration. For instance, one might argue that there are many beings, but there is one adequate theory about it, thus tempering ontological pluralism with theoretical monism.[47] Along other lines, one might argue that that there are many beings, but they can become one through a process of creativity, where the emergence of a novel entity is occasioned by the many becoming one.[48]

Is there a pluralism that affirms the plurality of irreducibly other, strange singularities of coexistence? This is a crucial question for coexistentialism. In Nancy's post-Heideggerian sense of things, this question touches on the "singular plural *with*," "the singular plural *co-*" of coexistence, "being-with-being-itself."[49] For Nancy, the arrival of this plurality coincides with the deconstruction of Christianity, opening up the foundations of monotheism and of the secular rationalism and individualism that replaced it, and thereby making way for practices of celebrating and adoring the plurality of coexistence.[50] In the spacing opened with the event of deconstruction, the plurality of things can arrive. They do not have meaning or make meaning. They do not make sense or have a sense. Things are their meaning. They do not have a sense. They are sense. In that sense, in Heidegger's rejection of nonhuman openness to the world, he is partially right to say that a stone is without world. He is wrong to deny agency to the stone, as it touches and is touched by other things. He is right that a "stone does not 'have' a sense," for it is sense, sharing in the world without having access to it.[51]

Nancy is careful to distinguish this kind of pluralism from "an animism or a panpsychism," which endow "the stone with an interiority."[52] Nancy's pluralism is attempting to celebrate the irreducibly singular differences constitutive of every being in the circulating plurality of beings. Injecting interiority into beings effaces their singularity. However, there are two problems in Nancy's account. First of all, he admits that he still has not found a way to articulate the differences between beings. Each being is a singularity, but a cow singularity is not the same as a candle, a crustacean, capitalism, or a college campus. For Nancy, "a 'quantum philosophy of nature' (or an 'atomistic' or 'discrete' one) remains to be thought."[53] This is not just Nancy's own shortcoming. Throughout the history of philosophy, philosophers have tended toward reducing the specific differences between things to fluctuating manifestations of some underlying substratum (matter, nature, process), to human projections (social construction, language games, nominalism), or caught somewhere in between, as in form/matter dualisms. The second problem in Nancy's account is related to the first. One of the reasons that Nancy—again, along with most of the European tradition of philosophy—has not thought through a quantum philosophy of nature is that he has misunderstood animism, which can provide a pluralism that celebrates the irreducible singularities and discrete differences that make up the mesh of coexistence.

Animism is far from an innocent term, with origins in anthropological discourses complicit in colonialism, cultural appropriation, and ontotheological assumptions about God, humans, and nature. Its resurgence through the work of Graham Harvey has helped establish a sense of animism that faces the disastrous legacies of anthropology while also affirming the usefulness of the word to describe a sense of the world found among indigenous lifeways and some environmentalists, philosophers, and adherents of Paganism. It is a sense of the world for which personhood or soul (*anima*) is not the sole property of humans.[54] Personhood is distributed across all beings, more or less, not in any homogenized or calculable totality. Animism does not theorize any pan of souls in the way that panpsychism does. A tree is different than a stone, which is different from a wolf or a mountain. They are not different manifestations of one archetypal complex called "soul." They are not one or different aspects of the One. "Animism is neither monist nor dualist, it is only just beginning when you get beyond counting one, two. . . . At its best it is thoroughly, gloriously, unashamedly, rampantly pluralist."[55] This pluralism is without interiority. Things do not have interiority, subjectivity, or personality. They are persons. "The world is full of persons (people if you prefer), but few of them are human."[56] Persons are unassimilable others. In the relationships of ecological existence, persons are unassimilable others in intimate contact with one another. They are "intimate alterities," not entities with an added layer called interiority or meaning.[57] One of the ways that animists avoid extraneously adding layers of subjectivity or sense inside or on top of things is through ecocritical performances that exceed the limits of academic tractates and manifestos. "Animism is better communicated in trickster tales, soulful songs, powerful poems, rousing rituals, and/or elemental etiquette than in manifestos."[58] Those ecocritical performances are about respecting persons, which is not to say that they are either descriptive or prescriptive. Those performances are ways of adoring the plurality of persons. They are not describing identities or prescribing norms for ethical action. Trickster tales disturb the identities and norms that efface the intimate alterities of persons. Animism is not descriptive or prescriptive. It is a way of attuning to irreducibly different modes of planetary coexistence. Accordingly, it is taken up in the ecological existentialism of Deborah Bird Rose.

In Rose's ethnographic research of Aboriginal people in Australia, she elaborates on the ways of being in the world performed in Aboriginal experiences of "country"—the place, home, or ecological community in and as which persons relate with one another. Rose describes country using a term from Levinas, "nourishing terrain": Country is a place that gives and receives life. Not just imaged or represented, it is lived in and lived with. . . . Country is home, and peace; nourishment for body, mind, and spirit; heart's ease."[59] Country is a person. You can speak to it, sing to it, visit it, desire it, care for it, and so on. Country is a person full of persons: trees, humans, dingoes, rivers, kangaroos, and stones. Persons are never alone, not even humans. Persons are in multispecies kinship groups, and those kinship groups have ancestral lineages that originate in "Dreamings," shapeshifting "creator beings" whose transformative tracks crisscross the country, showing signs of the events that led to the formation of kinship groups, such as the clan that includes kangaroo humans and kangaroo animals, or dingo humans and dingo animals.[60] Related in a family or clan, relatives take care of one another. Dingoes regard humans, and humans regard their dingo kin.

By remembering the Dreaming through trickster tales, songs, rituals, and poems, humans attune to their kin and respond to their intimate alterities. Along those lines, this "way of being in the world is a form of animism," specifically as animism is "defined in a new and excellent study by Graham Harvey," which is to say, animism as a recognition that "the world is full of persons."[61] Moreover, Rose also finds this animism in the ecofeminist philosophy of Val Plumwood, who affirms a complex sense of continuity and difference with human and Earth others.[62] In his ethnography of the Runa people of Ecuador's Upper Amazon, Eduardo Kohn describes a similar sense of animism as an attunement to continuity and difference:

> Runa animism is pragmatically oriented. The challenge for the Runa, as people who engage intimately with the beings of the forest in order, in large part, to eat them, is to find ways to enter this vast ecology of selves to harness some of its plenty. This requires being attuned to the unexpected affinities we share with other selves while at the same time recognizing the differences that distinguish the many kinds of selves that people the forest.[63]

Animistic kinship with alterities opens up ethical responsibility to the swarm of coexistence, face-to-face with the challenges of multispecies living and dying. Studying indigenous lifeways, Rose is particularly aware that decolonization must be included in this animistic sense of ethical responsibility.[64] If she was going postal, she might describe her form of animism as a postcolonial, posthumanist, postnatural, postsecularism. She calls it ecological existentialism.

The ecological existentialism of Rose's multispecies ethnography intersects with the coexistentialism of Morton's object-oriented ontology. For both of them, this ecologically updated from of existentialism is an ethical orientation of respect for the compelling alterity of other humans and nonhumans.[65] For both of them, the ways of being in the world that they represent can be considered forms of animism, as Morton puts it, "an upgraded form of animism," one that is attuned to the unique challenges of our current ecological emergency.[66] However, the term "animism" is tricky. It is important to keep in mind that this animism does not presuppose an opposition between life and death, animate and inanimate. While it is useful "to inject a little bit of animism into the discussion," it would be preferable to use a term that neither reduces life to nonlife (mechanism) nor reduces nonlife to life (vitalism), a term such as "*undead*."[67] This is a hauntingly strange animism, if it is animism at all.

To warn against mistaking this animism for a "belief system," for a perspective that opposes the living and nonliving, or for an attempt to unlearn modernity and return to precolonial, premodern animism, Morton crosses out the term: "~~animism~~." In the parlance of deconstruction, it is animism under erasure (*animisme sous rature*).[68] Under erasure, it conveys different ways of attuning to the intimate alterities of beings. Along those lines, an atheist, Muslim, Christian, agnostic, and a Buddhist can all practice animism. If you care about things, you might be an animist, even if one of the things you care about is God. In this context, God would not be understood as an omnipotent superbeing or causeless cause but more like a deconstructive event, such that "God" is a name that harbors a compelling call for justice, which can be heard issuing insistently from the alterity of every other.[69] In an "object-oriented theology," love of God would not be inconsistent with an animistic concern for intimate alterities.[70]

Morton is not deriving his animism from studies of indigenous lifeways but from his work with ecological philosophy, specifically the articulation of a "dark ecology" that affirms—with "a note of existentialism"—the paradoxes, uncertainties, and profound ambiguities of ecological coexistence without nature, leaving behind the "beautiful soul syndrome" that sees only reconcilable differences and seeks unity through embeddedness and nature immersion.[71] Without nature, there is just the swirling dust of singularities, intimate alterities, or as Morton says, "strange strangers," and the swirl of that dust is the "mesh," the complex network of interconnectedness between the hordes of strange strangers.[72] To think ecologically is to attune with the exigencies of radical coexistence, responding to the interconnectedness of things and their strangely compelling singularity. In short, to think ecologically is to face intimate alterities. The twofold thought of the mesh and strange strangers parallels in some ways Rose's characterization of ecological existentialism in terms of two shifts that define the current moment of planetary coexistence: "the shift into uncertainty and the shift into connectivity."[73] The shift into connectivity is the shift into the mesh, the intimate intertwining of human beings with one another and with the myriad denizens of Earth. This shift has come about with the articulation of ecological and evolutionary facts of interconnectedness and also with processes of globalization, which have intensified connections between humans and nonhumans across the planet. Furthermore, the more we humans experience this intensifying connectivity, the more apparent it is that what we are experiencing is profoundly uncertain—ambiguous, unpredictable, and dreadfully strange. Intensifying uncertainty can be found in scientific discoveries regarding the complex, nonlinear, unpredictable (stochastic) dynamics of material and organic systems. More horrifying, this uncertainty radiates from the proliferation of unpredictable events occasioned by massively distributed entities ("hyperobjects") like global warming, capitalism, and the current mass extinction event.[74] Those are unbearably intimate alterities to which you and I are called to respond.

The animism of ecological existentialism is out of step with modernity, specifically the situation of the modern subject, which pretends to be an all-powerful agent in control of a passive, mechanistic nature, along with countless underprivileged humans caught on the nature side of power. A troubling trend in the history of the moderns, as Latour puts it, "is not that there are still people 'mad enough to believe in animism,' but that so many hardheaded thinkers have invented what should be called *inanimism*," for which the agency of nonhumans becomes "an agency without agency constantly denied by practice."[75] Inanimism involves "a profound form of impoliteness toward nonhumans, who were said to be incapable of making a difference," who were restricted to a poor or deprived openness to the world, a "stupid being-there."[76] Haunted by the recalcitrant agency of nonhumans, inanimism could be called exanimism. It is a ghastly exhumation, an examination of the dead (*examinis de exanimis*). Yet the swarms of the so-called dead are not simply lifeless. Undead, the swarms swoon (*examina exanimata*), distressed and distraught (*exanimat*).

Modernity has been the site of a maddening proliferation of hybrid actors that cross boundaries between the supposedly separate realms of nature and culture, yet it is precisely in that context that nonhuman actors are most vehemently denied any regard and the nature/culture dualism is most robustly erected. The irony is that, while nonhuman agents are excluded and deprived of agency, the supposedly powerful human has become increasingly passive through regular immersion in the cozy comforts of modern pampering—phones, video games, computers, cars, and other poor replacements for the warming spheres of God and nature. An "animism in step with modernity," as Sloterdijk observes, is an animism for "sphere-deficient private persons," who are themselves immobile—relatively inanimate—while they "stare out through the media window into moving landscapes of images," such that "the moving images have become far livelier than most of their observers."[77] Supposedly passive nonhumans exhibit far more agency than the infantilized, media-addicted humans that live off of the systematic exploitation of nonhumans. A phone is sometimes more interesting than the human using it.

This is not to say that a coexistentialist animism has all the answers. Quite the opposite, this animism is under erasure, conveying an affirmation of the plurality of beings while holding in abeyance any determinate animism that could provide a framework or model to guide life on Earth toward life instead of death. This animism without animism opens out of the deconstructive event announced in the death knells for God and nature, freeing adherents of any or no religions to attune to the struggles of ecological coexistence, and freeing environmental scientists from the residual inanimism of modernity. Animism *sous rature* animates (without prescribing) new possibilities for ecological perspectives to respond to the unfathomable and irreducible strangeness of things.

As in Thacker's darkness mysticism, the climatological scale is particularly relevant for this animism. With the collapse of God and nature came the collapse of any psychocosmic immune system, including that of the *anima mundi*. After God and nature, there is no soul of the world. As humans become aware of ecological emergency, we are becoming aware of the importance of the climatological immune system of planetary coexistence. In an era of global warming, humans are "facing Gaia."[78] The dust of this planet haunts the remains of the *anima mundi*. The development of Earth systems sciences (Gaia theory) and advent of global warming call for an updated *anima mundi*.[79] Aware of the uncanny pulse of planetary coexistence, *animisme sans animisme* is not certain about any soul, world, or uncertainty. More ecognostic than agnostic, this animism is about remaining open to the plurality of intimate alterities, attuning to human and nonhuman persons. There is no soul that belongs to the world or to Earth. Earth does not have soul, is not soulful or ensouled.

Earth is a person. Like Aboriginal country, it is a nourishing terrain supporting the coexistence of other persons, agents, strange strangers. Currently, those strangers are caught up in multitudinous swarms of the distraught and distressed. The vulnerable exposure of those persons in their cohabitation remains to be thought. The next two chapters elaborate on the compelling challenges of inhabiting the nourishing terrains of a hauntingly animate planet, being with others, being in terrains.

NOTES

1. Ken Kolsbun, *Peace: The Biography of a Symbol* (Washington, DC: National Geographic Books, 2008).
2. Heidegger, *Contributions to Philosophy (From Enowning)*, 75. I discuss this in the beginning of the previous chapter ("Existentialist Legacies").
3. Gilles Deleuze, *Difference and Repetition*, trans. Paul Patton (New York: Columbia University Press, 1994), 52.
4. For a comprehensive overview of the field of religion and ecology, including an account of its place within the environmental humanities, see Willis Jenkins, Mary Evelyn Tucker, and John Grim, eds., *Routledge Handbook of Religion and Ecology* (New York: Routledge, 2016).
5. In the queer, paradoxical fashion of coexistentialism, out is in. I discuss this in the concluding section of the previous chapter ("Existentialist Legacies").
6. This is a more complex sense of "after" than the "after nature" of Jedidiah Purdy, who neither rethinks the idea of "nature" or the idea of being "after"; rather, he discusses the history of human-nature interactions to reflect on the political challenges facing our current epoch, the Anthropocene, in which anthropogenic effects have entangled humans with nature on a planetary scale. Jedidiah Purdy, *After Nature: A Politics for the Anthropocene* (Cambridge: Harvard University Press, 2015).
7. Mark C. Taylor, *After God* (Chicago: University of Chicago Press, 2009), 345.
8. Nietzsche, *The Gay Science*, 181 (#125).
9. Martin Buber, *Eclipse of God: Studies in the Relation Between Religion and Philosophy*, trans. Robert Seltzer (New York: Humanity Books, 1988).
10. Martin Buber, *I and Thou*, trans. Walter Kaufmann (New York: Touchstone, 1970), 56.
11. Buber, *Eclipse of God*, 23.
12. Ibid., 129.
13. Michel Foucault, *The Order of Things: An Archaeology of the Human Sciences*, trans. Alan Sheridan (London: Routledge, 2004), 422.
14. Nietzsche, *Twilight of the Idols*, 21.
15. Ibid., 24.
16. Peter Sloterdijk, *Globes: Spheres Volume II: Macrospherology*, trans. Wieland Hoban (Los Angeles: Semiotext(e), 2014), 449.
17. Bill McKibben, *The End of Nature* (New York: Random House, 2006). Carolyn Merchant, *The Death of Nature: Women, Ecology, and the Scientific Revolution* (San Francisco: HarperSanFrancisco, 1990).
18. Morton, *Ecology without Nature*.
19. The undecidable "whot" is a term that comes from John Llewelyn in his deconstructive inquiry regarding the undecidability of the who or what of the other, including divine, human, living, and inanimate others. John Llewelyn, *The Rigor of a Certain Inhumanity: Toward a Wider Suffrage* (Bloomington: Indiana University Press, 2012), 197.
20. Michael Marder, *Dust* (New York: Bloomsbury Academic, 2016), 35.
21. Ibid., xi.

22. Graham Harman, *Guerilla Metaphysics: Phenomenology and the Carpentry of Things* (Chicago: Open Court, 2005), 251.
23. In the words of Blake's "Auguries of Innocence" (1803), each human has the ability to see infinity in all things: "To see a World in a Grain of Sand / And a Heaven in a Wild Flower / Hold Infinity in the palm of your hand / And Eternity in an hour." William Blake, *The Complete Prose and Poetry of William Blake*, rev. ed., ed. David V. Erdman (Berkeley: University of California Press, 2008), 490.
24. Nancy, *Being Singular Plural*, 9.
25. Latour, *Politics of Nature: How to Bring the Sciences into Democracy*, trans. Catherine Porter (Cambridge: Harvard University Press, 2004), 25–26.
26. David Webster, *Dispirited: How Spirituality Makes Us Stupid, Selfish, and Unhappy* (Winchester: Zero Books, 2012), 1.
27. Pierre Hadot, *Philosophy as a Way of Life: Spiritual Exercises from Socrates to Foucault*, trans. Michael Chase (Malden: Blackwell Publishing, 1995).
28. Webster, *Dispirited*, 17.
29. Ibid., 7.
30. Ibid., 71.
31. Ibid., 75.
32. Ibid., 40.
33. Ibid., 76.
34. Ibid., 71.
35. Stuart Kauffman, *At Home in the Universe: The Search for Laws of Self-Organization and Complexity* (Oxford: Oxford University Press, 1995).
36. Whitney A. Bauman, *Religion and Ecology: Developing a Planetary Ethic* (New York: Columbia University Press, 2014), 74.
37. Morton, *Ecology without Nature*, 184.
38. Nietzsche, *The Gay Science*, 168 (#109).
39. Rose, *Wild Dog Dreaming*, 47.
40. This term comes from Timothy Morton, *Dark Ecology: Toward a Logic of Future Coexistence* (New York: Columbia University Press, 2016). This book derives from his lectures for the 2014 Wellek Library Lecture series hosted by the Critical Theory Institute at the University of California, Irvine. For a short time, Morton was using the title of "Ecognosis" instead of "Dark Ecology" for the lectures. Timothy Morton, "Wellek Lectures: The Description," *Ecology without Nature* blog, April 18, 2014. ecologywithoutnature.blogspot.com/2014/04/wellek-lectures-description.html?utm_source=twitterfeed&utm_medium=twitter.
41. Morton, *Dark Ecology*, 5.
42. The definition of religion in terms of training is explored in chapter 10 ("*Askesis*: Shut Up and Train!").
43. Marder, *Dust*, 80.
44. Thacker, *In the Dust of This Planet*, 6.
45. Ibid., 8–9.
46. Ibid., 158–59.
47. This is what the environmental philosopher J. Baird Callicott proposes in his presentation of an Earth ethic based upon the land ethic of the American conservationist Aldo Leopold. J. Baird Callicott, *Thinking Like a Planet: The Land Ethic and the Earth Ethic* (New York: Oxford University Press, 2013), 12.
48. This is how Alfred North Whitehead defines creativity. It is "the principle of novelty" whereby the plurality of the universe becomes unified and added to: "The many become one, and are increased by one." Alfred North Whitehead, *Process and Reality: An Essay in Cosmology*, ed. David Ray Griffin and Donald W. Sherburne (New York: Free Press, 1978), 21.
49. Nancy, *Being Singular Plural*, 38.
50. See Nancy's two-volume work on the deconstruction of Christianity. Jean-Luc Nancy, *Dis-Enclosure: The Deconstruction of Christianity*, trans. Bettina Bergo, Gabriel Malenfant, and Michael B. Smith (New York: Fordham University Press, 2008). Jean-

Luc Nancy, *Adoration: The Deconstruction of Christianity II*, trans. John McKeane (New York: Fordham University Press, 2013).

51. Jean-Luc Nancy, *The Sense of the World*, trans. Jeffrey Librett (Minneapolis: University of Minnesota Press, 1997), 63.

52. Ibid., 62.

53. Ibid.

54. Graham Harvey, *Animism: Respect for the Living World* (New York: Columbia University Press, 2005). Graham Harvey, ed. *The Handbook of Contemporary Animism* (New York: Routledge, 2014).

55. Graham Harvey, "An Animist Manifesto," *PAN: Philosophy, Activism, Nature* 9 (2012), 3.

56. Ibid., 2.

57. Ibid.

58. Ibid., 4.

59. Deborah Bird Rose, *Nourishing Terrains: Australian Aboriginal Views of Landscape and Wilderness* (Canberra: Australian Heritage Commission, 1996), 7. Rendered as "nourishing terrain," Levinas uses the French "*espace vital*," literally "living space," like the German "*Lebensraum*," in the sense of a territory necessary for the well-being of a people. For Levinas, even a book can become a nourishing terrain, as in the case of the Bible for the people of Israel, "the people of the Book." Emmanuel Levinas, "Revelation in the Jewish Tradition," in *The Levinas Reader*, ed. Sean Hand (Oxford: Blackwell Publishers, 1997), 192, 210n2.

60. Rose, *Wild Dog Dreaming*, 17.

61. Ibid., 18.

62. Deborah Bird Rose, "Val Plumwood's Philosophical Animism: Attentive Interactions in the Sentient World," *Environmental Humanities* 3 (2013). Plumwood appreciates animism while also warning against "dualising forms of animism," which "can result from acceptance of the dualising of nature and spirit, and the addition of spirit as an extra ingredient needed to make nature spiritual or subjective." Val Plumwood, *Feminism and the Mastery of Nature* (London: Routledge, 1993), 136.

63. Eduardo Kohn, *How Forests Think: Toward an Anthropology Beyond the Human* (Berkeley: University of California Press, 2013), 95. Kohn refers to the works of Philippe Descola and Eduardo Viveiros de Castro as examples of theorists who are working with the term "animism" that critiques Western conceptions of "nature" while also avoiding evolutionist and racist connotations of animism related to the use of the term in the early twentieth century (ibid., 93). Philippe Descola, *Beyond Nature and Culture*, trans. Janet Lloyd (Chicago: University of Chicago Press, 2005); Eduardo Viveiros de Castro, *The Relative Native: Essays on Indigenous Conceptual Worlds* (Chicago: Hau Books, 2015).

64. Deborah Bird Rose, *Reports from a Wild Country: Ethics for Decolonisation* (Sydney: University of New South Wales, 2004).

65. "The ecological thought needs to develop an ethical attitude we might call 'coexistentialism.'" Morton, *The Ecological Thought*, 47.

66. Ibid., 110.

67. Timothy Morton, *Realist Magic: Objects, Ontology, Causality* (Ann Arbor: Open Humanities Press, 2013), 101.

68. Morton, *Hyperobjects*, 172.

69. Regarding God as a deconstructive event harbored in the name "God," calling for responsibility to every other as wholly other, see John Caputo, *The Weakness of God: A Theology of the Event* (Bloomington: Indiana University Press, 2006); John Caputo, *The Insistence of God: A Theology of Perhaps* (Bloomington: Indiana University Press, 2013).

70. Adam Miller, *Speculative Grace: Bruno Latour and Object-Oriented Theology* (New York: Fordham University Press, 2013).

71. Morton, *Ecology without Nature*, 143, 181–88. Cf. Morton, *Dark Ecology*.

72. Morton, *The Ecological Thought*, 15, 28–50.

73. Rose, *Wild Dog Dreaming*, 42.

74. Morton "coined the term hyperobjects to refer to things that are massively distributed in time and space relative to humans." Examples include "a black hole," "the Lago Agrio oil field in Ecuador," "the Florida Everglades," "the biosphere," "Styrofoam," "capitalism," etc. Morton, *Hyperobjects*, 1.

75. Bruno Latour, "An Attempt at a 'Compositionist Manifesto,'" *New Literary History* 41 (2010), 482–83.

76. Latour, *Politics of Nature*, 283n33. For more on Latour's conceptions of animism and inanimism, see Bruno Latour, "Angels without Wings: A Conversation Between Bruno Latour and Anselm Franke," in *Animism: Volume 1*, ed. Anselm Franke (Berlin: Sternberg Press, 2011).

77. Sloterdijk, *Bubbles*, 73.

78. Bruno Latour, *Facing Gaia: Six Lectures on the Political Theology of Nature*. Gifford Lectures on Natural Religion. Edinburgh, February 18–28, 2013.

79. "This dawning awareness of the *anima mundi* in our times is in truth a reawakening of the old, non-dualistic animism that has been dormant so long. . . . Our task now is to explore ways in which the new animism can be integrated into the very heart of Western culture." Stephan Harding, *Animate Earth: Science, Intuition, and Gaia* (White River Junction, VT: Chelsea Green Publishing Company, 2006), 29.

FOUR

Remaining Exposed

Bloody, beaten, and beating, and somehow still demanding further beating, pounding, striking, striking-together, incessant syncopation, ongoing syncope, a pulsating impulse that is always passing out, transitioning between contraction and expansion, systole and diastole—the heart. Is anything or anyone closer to you than your own heart? A heart is intimately vulnerable, exposed to indeterminable others. It is exposed to intruders and guests that could destroy the heart in a variety of ways, as is evident in a spectrum of examples ranging from biomedical (e.g., cardiac arrest, heart disease) to moral-psychological (e.g., heartlessness, coldheartedness), which seem to blend in cases like stress cardiomyopathy, the broken heart syndrome wherein emotional or physical stress causes a severe, rapid, and potentially life-threatening weakening of the heart muscle. The exposure of a heart to its surrounding conditions allows the heart to function, taking in and redistributing blood, yet that same exposure makes the heart susceptible to intruders and irregularities. The heart discloses a coexistential aporia: exposure or some such vulnerable openness is characteristic of the constitution of things as well as their destruction, rendering each body possible and impossible. It is sometimes difficult to know how much is too much, as instances of sunburn, hypothermia, indecent exposure, and media saturation all indicate.

Attending to the dynamics of the heart's intimate vulnerability, one can learn about much more than the heart. Thinking of the heart, taking it to heart, can one get to the very heart of things? Thinking is perhaps at its best when it is a thinking that takes to heart, thinking (*denken*) as thanking (*danken*). In Heidegger's meditative etymological exercises, thinking and thanking come together at their root in "*thanc*" (Old English; "*Gedanc*," Old German), which he describes as "the heart's core"—"the gathering of all that concerns us, all that we care for, all that touches us insofar as we are, as human beings."[1] That concluding reference to human beings is indicative of the anthropocentric tendencies in Heidegger's thinking. What about all that touches nonhuman beings? Only humans think, care, and take to heart, in the same way that Heidegger argues that only humans are "world-forming" (*weltbilden*), with animals relegated to the status of "poor in world" (*weltarm*) and stones completely left out in the cold, "worldless" (*weltlos*).[2] For Heidegger, animals must be poor in heart, and a stone is heartless. Of course, a stone does not have a heart in the anatomical sense, but it does have a heart (*cor*) insofar as it has a core, which could be understood as a physical core—the middle of the stone, a depth beneath its surface. The core of a stone could also be thought ontologically as the singularity of the stone, which is not a particular layer of the stone but is the stone's sheer existence, being itself. Inexhaustible, unfathomable, and ineffable, that singularity cannot be obtained. You cannot get blood from a stone. In that sense, its heart can be thought of as a depth beneath the surface appearances of the stone, the unobtainable core beneath appearances that can be calculated, measured, and spoken about. However, that metaphor is misleading, as the singularity of the stone is not some added layer or injected interiority.

The stone wears its heart on its sleeve. Its singularity is the singularity of its exposure, its sheer nudity. You can talk about it, kick it, lick it, take a photograph of it, contemplate its existence, and destroy it. The alterity of the stone is not behind the scenes. The real stone in all of its singular strangeness is right here, in your face. Yet, as long as it exists, there is always a surplus of its being, a surplus of alterity, no matter what I think about the stone or do to it, no matter what it means or does for the others with which it comes into contact. For Levinas, this is the exorbitance of the nakedness of things.

> They are always in some respect like those industrial cities where everything is adapted to a goal of production, but which, full of smoke, full of wastes and sadness, exist also for themselves. For a thing nudity is the surplus of its being over its finality. It is its absurdity, its uselessness.[3]

As Morton observes, this is indicative of a coexistentialist element in Levinas, as it conveys the paradox of the mesh (connectivity) and strange strangers (alterity).[4] A thing is irreducibly itself, yet it is exposed to networks of connections and processes relating to other things. This is not simply to say that a thing is actively itself while passively receiving the contact of other things. The activity of a thing and its receptivity to others are both irreducible to the nakedness of the thing itself. That relative activity and passivity open out of the radical passivity of exposure, the absurd and useless passivity of being-with.[5]

Naked, beings are themselves, yet their exposure reduces them to what other beings make of them. Coexistence means being nude, vulnerable, exposed to others who are themselves nude, vulnerable, exposed. In Nancy's terminology, coexistence is "being-with-being-itself."[6] I am myself and not myself, irreducibly unique yet intimately interwoven with others whose contact translates my existence into their horizons. I am enmeshed with strange strangers. I am grasped by others, yet those others remain intact, ungraspable by me while I am nonetheless in contact with them. If I take a walk down a path, and I see a steep hill far ahead, my eyes make contact with the hill, and I am grasped by the hill. Whether I decide to continue on the path toward it or avoid the hill by walking back or finding a way around, I cannot avoid relating to the hill. Exposed, I am caught by the hill, this hill, in its obstinate singularity, transformed by the compelling light of its alterity. This is like the hill of which Rilke speaks in his poem, "A Walk" (*Spaziergang*). His eyes "already touch the sunny hill" in the distance ahead of the path he started on, and "even from a distance" he is caught and changed by its ineffable "inner light": "So we are grasped by what we cannot grasp."[7] When we are touched by the untouchable, we are changed into something different, which is something that, as Rilke points out, "we already are." I become who I am by ceasing to become who I am. Being-with is the core of my being. Being outside of myself is closer to me than I am to myself.

Remaining exposed, I am inside out, possible and impossible. This is not unique to me or to humans but is constitutive of any entity. To be is to be with, to wear your heart on your sleeve, to expose a surplus of singularity to others. If the heart exemplifies an exposure or exposition characteristic of all beings, attention to the vulnerability of the heart could facilitate an understanding of the role of vulnerability in intimate coexistence in general. In this chapter, the vulnerability of the heart opens onto the vulnerability of planetary coexistence during a time of increasingly intimate intertwining of human bodies with one another and with the other inhabitants and habitats of Earth. Perhaps it is possible today to join in Augustine's confession that "our hearts are restless" (*inquietum est cor nostrum*), although not in order to find solace in the transcendent God of monotheism but in order to affirm the restless swarm of coexistence during a time of ecological emergency.[8]

To analyze the intimate vulnerability of exposure, I draw on Sloterdijk, whose writings convey a coexistentialist affirmation of intimacy, articulating the being-with of beings through a mutant mode of metaphysical speculation—a General Immunology or General Spherology—which resuscitates the abandoned task of metaphysics in the wake of post-Kantian philosophies of consciousness, critique, and communicative action. Sloterdijk's work shares an impulse with the heart of another contemporary philosopher, Jean-Luc Nancy, whose experience of undergoing a heart transplant has profoundly informed his corpus. Sloterdijk and Nancy both follow Heidegger's conception of being-in-the-world while refuting the loneliness that characterizes human existence for Heidegger. Another thinker who develops notions of vulnerability while critiquing Heidegger's poor account of world is the feminist theorist of science Donna Haraway, whose work companions well with the ecological existentialism of Deborah Bird Rose.[9] Following the heart of each thinker, there are two pulses to consider: first, the contracted pulse, systole, the micro-scale of one-on-one relations, and second, diastole, expanding to the macro-scale of planetary interconnectedness.

SYSTOLE

Whence the preoccupation with hearts, the injunction to be cordial, hearty, and heartfelt, in touch with the heartland, warmhearted, openhearted? It is not new. Rudimentary heart-shaped symbols can be found among the earliest cave drawings.

> The heart symbol was used in a multitude of cultures: in Ancient Egypt, it symbolized the centre of life and morality; in Ancient Greece it was seen as the centre of the soul and as the source of heat within the body; the Romans were the first to understand that the heart is the most vital organ in sustaining life. The symbol became even more widespread from the Middle Ages onwards.[10]

The ubiquity of heart symbolism should not be mistaken for universality. Although all humans and indeed all animals have hearts, not all cultures designate the heart as the site of one's innermost self or one's emotional or spiritual life. There are similar designations across cultures, although not focusing on the heart per se. For instance, there are Japanese terms for one's central emotional sphere, but they are not focused specifically or exclusively on the heart; *kokoro* can refer to "heart" as well as "sense" or "spirit," and *hara* refers to the "belly" or the "center of the body."[11] Similarly, the heart chakra of yogic traditions is not exactly cardiocentric. It is the *anahata* chakra. *Anahata* means "unstruck" or "unbeaten," and although it refers to an energy channel situated along the spine in the center of the body near the heart, it is not exclusively related to the heart but related also to silence, skin, touch, and air. Far from universal, the equation of the heart with one's central self, one's core identity, emerged within the context of Christian discourses and disciplines. That is the beginning of Sloterdijk's analysis of the heart.

"Cordial subjectivity" was the norm for "Christianized Europeans, especially in the Middle Ages and early Modern Age."[12] The crucial characteristic of cordial subjectivity is the pathology if not impossibility of holding onto one's own heart. Hearty subjectivity is constituted in heart-to-heart relationships ("bipolar cardial relationships"), of which Sloterdijk gives multiple examples, including two from the fourteenth century: a novella by the poet Konrad of Würzburg and an event in the life of

Catherine of Siena, an Italian mystic.[13] A brief consideration of those examples can provide a sense of the vulnerable intimacy of the heart.

Konrad's novella, *Herzmaere* ("Heart Fable"), presents a romantic tale of courtly love. In the age of chivalry, a knight and his lady are so intimately in love that they become one, but the lady is already married, so their noble love is doomed. The lovers become separated, the knight having only a ring from the lady as a token of remembrance. While the knight is living a lonely life in another country, he dies, having previously instructed his squire to carry out a series of actions: to cut the heart out of the knight's dead body and deliver it to the lady along with the ring to assure her that it is truly the knight's heart. When the squire arrives, the lady's husband confronts him, and upon learning of the squire's mission, the husband has his cook prepare a meal with it, which is then served to his wife. After she consumes the meal, which is most delicious, her husband explains that she just ate the knight's heart, at which point her heart freezes, blood comes out of her mouth, and her heart breaks. Frozen and broken, her heart repeats the noble deed of the knight, demonstrating a shared devotion considered exemplary for courtly love.

Subjectivity in this context is realized as an intersubjective intertwining, an exchange of hearts consummated in consumption. That cordial subjectivity is a worldly transposition of Christian mysticism and its metaphysics of unity through communion. Here is my body; here is my heart (*Hoc est corpus meum; hoc est cor meum*). With transubstantiation taking place in the kitchen instead of the altar, the knight's heart replaces the Eucharist—the ritually consumed body of the incarnation of love, Jesus Christ. Although it points to a mystical unity that is presumably edifying, the anthropophagous dimension of cordial subjectivity is horrifyingly literal in the case of Konrad's novella, in contrast to its symbolic sublimation in Christian communion. For Sloterdijk, that cannibalistic horror "indicates how the path to biune intimacy in advanced civilization is hampered by mistaken embraces."[14] It is getting increasingly difficult to discern the appropriate entrance into the other. The horror of a literal transformation always haunts cordial subjectivity.

Catherine of Siena provides a more direct example of a heart-to-heart exchange. After praying for God to renew her spirit and create a "clean heart" within her, she had a vision of Jesus coming to her, opening her left side, taking out her heart, and leaving, and for some time following that event, she said she was "living without a heart," knowing that it is supposedly impossible to live under such conditions.[15] Eventually, another vision came to her while praying, and her Lord returned to her, opening up her left side and putting his heart in her body where her heart used to be. This heart operation is more direct than the heart exchange of Konrad's lovers, since Jesus takes out Catherine's heart and replaces it without mediation through eating. Furthermore, this heart operation is more extreme because it is not simply mixing the subjectivity of lover and beloved, but also of human and divine, such that cordial subjectivity mixes human subjectivity with a nonhuman subjectivity that is absolute—pure interiority completely withdrawn from anything external.

The examples from Konrad and Catherine indicate the profound intimacy of cordial subjectivity, an intimacy wherein it is impossible to hold onto one's own heart. The intercordial space is one among many intimate spaces that Sloterdijk examines, including the interfacial space, "magical" space of bonding (e.g., hypnosis, Mesmerism, transference), intrauterine space of mother-child relations and child-placenta relations (and their postnatal variants), the space of the care of one's own genius (or angel), and the resonant space of the welcoming voice (i.e., the maternal voice and its messianic variants).[16] The common thread in these intimate spaces is "inhood" (*Inheit*), which is Heidegger's term to indicate that the ecstatic openness of human existence—the "there" (*da*) of *Dasein*—"can only be *there* if it is contained, surrounded, encompassed, disclosed, breathed-upon, resounded-through, attuned and addressed."[17] Being-in or being-contained is thus a basic condition of human existence, which is not to say that humans live in physical or metaphysical containers but that human existence is suspended in concern, a "ball of care [*Sorge*] in which existence has spread out in an original being-outside-itself."[18] Being-in is the position of exposure, opening up to others, being with others and being outside of oneself.

Heidegger's sense of being-in undoes the physics and metaphysics of containers, such that humans are no longer conceived as rational or living things inside of a container called world, reason, God, nature, environment, etc. Human existence is carefully attuned to the being-outside-itself of existence, which is to say, human existence brings close the distance of things. "*In Dasein there lies an essential tendency toward closeness.*"[19] Bringing-close entails "*making room,*" giving things space.[20] Whereas Heidegger's analysis led to a focus on the loneliness and solitude characterizing the place of *Dasein* in the world, Sloterdijk is committed to refuting loneliness. Indeed, if a sign was posted at the entrance to Sloterdijk's *Spheres* trilogy, it would prohibit entrance to anyone who is unwilling "to refute loneliness," which is not entirely unlike the sign hanging at the entrance to Plato's Academy, "Let no one enter who is not a geometrician."[21] To refute loneliness, Sloterdijk practices something like "geometric vitalism."[22] He opens up the definition of human existence as being-in-the-world, exposing it to multiple worlds and multiple ways of being-in, outlining "multi-significant universes of existential spaciousness," for which Sloterdijk uses the geometric term "spheres."[23] In other words, "'being-in-the-world' first of all, and in most cases, means being-in-spheres."[24]

Spheres are not easy to define, as they are not a matter of access but only of circumspective concern. They are "the most real, yet also the most elusive and least tangible things."[25] Consider some possible definitions. Spheres are intimate spaces, which beings generate as they distinguish themselves from others, differentiating insides from outsides. "Spheres are immune-systemically effective space creations for ecstatic beings that are operated upon by the outside."[26] Spheres are "morpho-immunological constructs," or to put it another way, they are "air conditioning systems in whose construction and calibration, for those living in real coexistence, it is out of the question not to participate."[27] Spherology is another name for General Immunology. It accomplishes the metaphysical task of articulating the meaning of being, and it integrates post-Kantian critiques that challenge metaphysical protectionism and expose the conditions and agencies on which metaphysical systems are built. For Sloterdijk, "General Immunology is the legitimate successor of metaphysics."[28] Immunology is generalized to account not only for the dynamics of biological systems but all kinds of systems, including symbolic and conceptual systems and large-scale social systems. In other words, General Immunology is General Spherology. Its inquiries range from the microspheres of bipolar intimacy (i.e., bubbles) to macrospheres of collective coexistence (i.e., globes).

General Immunology studies the spheres appropriate for the flourishing of planetary coexistence, distinguishing between healthy spheres and instances of "sphere pathology," which are indicative of a failure to cultivate or maintain intimacy and solidarity.[29] If Konrad and Catherine provide examples of healthy intercordial spheres, an example of sphere pathology can be seen in the modern "anatomical disenchantment of the heart."[30] The modern period saw a rise in the previously disparaged practice of dissecting cadavers, giving rise to a new science of anatomy alongside a new anthropology. "The dissecting tables of the anatomists were transformed into the altars of the new science of humans; the corpses graduated as assistant lecturers in anthropology."[31] The main lesson of this new anthropology is "anatomical individualism": human bodies are primarily individual bodies that are unrelated to one another, participating in relationships only secondarily.[32] Such individualism bursts the bubbles of healthy microspheres by replacing the intimately vulnerable heart with a mechanistic heart, replacing sanguine warmth with cold machinery. Sacred heart theologies and Romantic resuscitations of hearty and heartfelt subjectivity resisted that cooling process, setting the stage for the ongoing "struggle for the calibration of the world's temperature."[33]

Anatomical naturalism is not entirely pathological. On one hand, the individualist lack of intimacy generates sphere pathology, but on the other hand, naturalistic or materialistic renderings of the heart emancipate the heart from its premodern metaphysical confines and expose it to the outside, opening it up to an exteriority that can bring greater opportunities for intense and transformative intimacy. To overcome its pathological aspects, the struggle to envision the world as a collection of machines must enter a new phase, such that machines are never considered isolated individuals but are generated and maintained "in bipolar, multipolar coexistence and an inter-intelligent bipolar circuit."[34] For an example of cordial intimacy with the exteriority of technoscientific machinery, consider the vulnerable intimacy of Jean-Luc Nancy's heart transplant.[35]

Severe health problems began for Nancy in the late 1980s, when he was in his forties. His "own heart, then, was useless for reasons never explained."[36] For years previously he had experienced flutters and irregular rhythms, and as symptoms worsened, Nancy was told by doctors that he needed a heart transplant. With an intersection between personal and technological contingency, Nancy received a heart transplant, undergoing a medical operation that first emerged only two decades prior. The heart operation is not the end of the struggle, but only the beginning. To welcome the heart of another, one must become exceptionally vulnerable, suppressing one's immunity to the intruding other. Accordingly, Nancy was given the drug cyclosporine to suppress his immune response and protect against the rejection of the new heart—the cordial stranger, grafted into him to replace his own heart, which was obviously no longer his "own" (if it ever was), not property, not properly functioning, already having become a stranger itself. Nancy's heart, having become intrusive, was extruded, and a new intruder was introduced, a stranger he was forced to welcome. The immunosuppression drugs made Nancy a stranger to himself, making him vulnerable not only to the transplanted heart but also to numerous unwanted intruders, including multiple bouts with cytomegalovirus (shingles), demonstrating a basic formula that marks the transition from metaphysics to coexistential immunology: "identity is equal to immunity."[37] After eight years of such problems, Nancy developed a cancer, a lymphoma, resulting in his forced hospitality to yet more hostile intruders: chemotherapeutic treatments, radiation, and the "brutalization and spaciness" that accompany the pain-relieving effects of morphine.[38] This demonstrates a point made by Sloterdijk: "It characterizes the basic risk of all intimacy that our destroyer sometimes gets closer to us than our ally."[39]

Like the examples from Konrad and Catherine, Nancy's example involves an exchange of hearts, although it is neither an exchange between a lover and the beloved nor an exchange between a mystic and the absolute. Nancy's heart was exchanged with that of a stranger, and the exchange was mediated not through love and not through divinity. It was mediated by the machinery of technoscientific instruments, drugs, procedures, and personnel. Yet, despite the context of cardiological disenchantment, Nancy does not describe his heart transplant in terms of anatomical individualism. He does not describe himself as an individual whose heart was exchanged for another heart and who contracted many illnesses in the process. His individuality is estranged from itself, exceeding itself and opening itself to relations with strange bodies, such that Nancy's "I" is not properly "his." I am and am not myself. I is and is not myself. Rather, "I end/s up being nothing more than a fine wire stretched from pain to pain and strangeness to strangeness."[40]

Nancy is not an individual "I" that contracted cancer after taking cyclosporine. His "I" is the cancer and the cyclosporine.

> I am the illness and the medicine, I am the cancerous cell and the grafted organ, I am these immuno-depressive agents and their palliatives, I am these ends of steel wire that brace my sternum and this injection site permanently sewn under my clavicle, altogether as if, already and besides, I were these screws in my thigh and this plate inside my groin.[41]

Those are not expressions of anatomical individualism but of multipolar intimacy, whereby Nancy's proper sense of self—his "I"—"plunges into an intimacy deeper than any interiority."[42] It is intimacy after the interior depths of nature, God, and the ersatz divinity of courtly love. It is intimacy with the outside, where subjectivity is exposed, shared out and distributed across bodies, organs, machines, intruders, strangers . . .

Such intimacy is not forged by reacting against disenchantment and resuscitating the hearts of Romanticism or premodern metaphysics. This intimacy is forged when I give my body to the other (Jesus: *Hoc est corpus meum*), opening myself up and letting the other come into the depths of my own heart, closer than I am to myself (Augustine: *interior intimo meo*). I am myself in my exposure as an exterior thing among things. The singularity of my being, closer to me than I can access, is my exorbitant being as a thing, my body, naked and vulnerable to the outside. "*Corpus meum* and *interior intimo meo*, the two being joined, in a complete configuration of the death of god, in order to say very precisely that the subject's truth is its exteriority and its excessiveness: its infinite exposition."[43] The cooling of the enchanted heart and the death of God thus mark a transition toward new ways of inhabiting spaces with intimate alterities, such that intimacy is not about creating an interior space over here, which is opposed to some exteriority over there or another interiority in there. After God and nature, intimacy becomes excessively vulnerable, remaining infinitely exposed, such that exteriority is found at the very heart of any interiority. As Nancy puts it, "the only 'outside' . . . is never the one we see out the window, which is only 'outside' due to its difference from the 'inside.' The real outside is not another inside than this inside: it is at the heart (you can say that again!) of the inside."[44]

DIASTOLE

Carrying out a transition from systole to diastole, from microspheres to macrospheres, from bubbles to globes, the question of intimacy becomes planetary. In particular, planetary coexistence has crossed a threshold of intimacy such that human activities, mediated with industrial technology and fossil-fuel energy, have occasioned the emergence of a new geological period, the Anthropocene epoch.[45] Humans, diversely distributed in terms of power, position, and privilege, have made noticeable and indeed life-threatening changes to Earth on an epochal scale. More than a river here and a forest there, humans have altered large-scale systems like the acidity of the oceans and the carbon dioxide in the atmosphere. Nuclear weapons testing has spread radioactivity around the planet. Plastic, Styrofoam, pollutants, and toxins have become ubiquitous, stamping the presence of humans across Earth's land, life, air, and water. The Anthropocene is an epoch of unprecedented intimacy, that is, the intimacy of planetary coexistence. Along those lines, Latour remarks in his lecture series *Facing Gaia* that Sloterdijk's vision of intimate spaces provides what could be considered "the first philosophy resonant with the Anthropocene."[46] Describing the *Spheres* trilogy—a "study of the envelopes indispensable for the furthering of life"—Latour notes that spherology is "part of a vastly expanded definition of *immunology,* understood here not as a human nor as a natural science but rather as the first *anthropocenic* discipline."[47]

To elucidate the anthropocenic implications of the vulnerable intimacy disclosed in heart-to-heart relationships, one must consider how "being-in on a small scale returns as a political and cosmic relationship."[48] Recall the modern disenchantment of the warmhearted world of Christianized Europe. Above the cold heart of anatomical individualism one finds the cold sky of post-Copernican astronomy, for which humanity dwells amid "monstrous externalities that breathe on it with stellar coldness and extra-human complexity," a coldness that renders untenable the idea that Earth is "enclosed by spherical forms like warming heavenly mantles."[49] Just as the cold heart discloses a loss of microspheric interiority, the cold sky conveys a loss of macrospheric interiority, which is to say, it conveys the fact that humans are destined to exist "only on a ball, but no longer inside a ball."[50] There is no existence under a transcendent heaven, only existence on an exposed Earth. After monotheism comes "monogeism," a shared commitment to the singular Earth.[51]

Not incidentally, the post-Copernican era of European history is one oriented toward the production of globes and various spherical images of Earth, providing some way of recapturing the warmth of the heavenly spheres, providing some consolation for the disenchantment of the world. While some were making globes, others were (and still are) wrapping themselves in worn-out illusions, trying to prevent or reverse the disenchantment process, as if they could hold up the cosmic roof while scientific dis-covery was tearing it down. In the meantime, the celestial domes collapsed, "the open animal tore down the roof of its old house from the inside."[52] This is what makes modernity unique: "after the turn to the Copernican world, the sky as an immune system was suddenly useless."[53] Having become useless, the immune systems developed by the traditional narratives of theology and cosmology needed to be replaced, and thus began the construction of the artificial spheres of globalization, including national governments, free markets, insurance policies, telecommunications, social media, and so on.

Modernity is a global heart transplant, a desperate effort to find a new source of vital warmth capable of replacing the lost macrospheric interiority. In other words, "modern humanity makes use of a deliberate greenhouse effect," opposing "the cosmic frost infiltrating the human sphere through the open windows of the Enlightenment" by creating an artificial hothouse for planetary coexistence:

> From this perspective, the Modern Age appears as the age of an oath sworn in offensive desperation: that a comprehensive house-building operation for the species and a policy of global warming must be successful faced with the open, cold and silent sky.[54]

The dominant mode of house-building today is anatomical individualism, where each individual is a rational consumer free to participate in the markets of global capitalism. In microspherological contexts, anatomical individualism results in isolated and disaffected humans who are not capable of developing their own emotional and cognitive spaces or building intimate relations with others. In macrospherological contexts, anatomical individualism entails a lack of political solidarity, such that there are no healthy macrospheres but only conglomerations of individual bodies, each in its own consumerist bubble.

Instead of a global civilization, there are foams of individual bubbles. "What is currently being confusedly proclaimed in all the media as *the* globalization of the world is, in morphological terms, the universalized war of foams."[55] The struggle for a global greenhouse is not working. Insofar as it does work, it only works for the very few who can afford a bubble of capitalist interiority, an individual share in the *Weltinnenraum* ("world interior") of capital, which is purchased at the expense of the vast majority, like "a comfort-animated artificial continent in the ocean of poverty."[56] In the war of foams, there is no chance for coexistence in a global community, indeed no real chance for being-in at all. The microspheres and macrospheres of interiority are gone. This is even true of the individual bubbles who can afford a comfortable share in the capitalist hothouse. All interiority is shown to be nothing but the functioning of exterior parts. Every interior is "increasingly exposed as the inner side of an outside," and "every blissful, intimate, vibrating cell is surrounded by swarms of professional disillusioners, and we drift among them—thought paparazzi, deconstructivists, interior deniers and cognitive scientists."[57] Through the immunosuppressive drug of disenchantment and the cancerous expansion of capitalist foam, humans are exposing all planetary systems to tremendous stress, "putting evolved immune systems at risk" and gathering all of the planet's inhabitants and habitats into "an ecological stress commune."[58]

How, then, can humans proceed with the task of constructing a global hothouse and finding some peace in the war of foams? How is it possible today to facilitate some sense of global being-in, some sense of intimate planetary coexistence, so that the Anthropocene does not continue following the lonely lines of anatomical individualism? A typical strategy for overcoming individualism and isolationism is inclusive community, but sphere theory renders untenable any dream of pure inclusivity or holism. An example of the horror of such radical inclusivity can be found in any totalitarian state, in the impacts of Christian missionaries throughout colonial modernization, or in any suppression and destruction of traditional micro- and macrospheres in the name of a universal community.

The task of participating in the struggles of planetary coexistence cannot escape the immunological imperatives of spherology. Every sphere, no matter how inclusive, must draw boundaries: inclusions and exclusion, intrusions and extrusions, friends and strangers, strengths and weaknesses. To fail in the production or maintenance of those boundaries is to fail to inhabit intimate space, which is ultimately a failure to exist. Vulnerable exposure to the outside is the very heart of the inside. The war of foams is too individualistic to cultivate intimate coexistence on microspheric or macrospheric scales, and collectivist alternatives to individualism tend to suppress otherness and homogenize differences. Indeed, "nothing misjudges the autonomous laws of micro- and macrospheres alike as much as the attempt to turn the whole dark, overpopulated world into a transparent and homogenous home for all without further ado."[59]

A global hothouse can be built neither on individualism nor on collectivism. It must be built on multipolar intimacies capable of remaining exposed, capable of ongoing exposure to the outside. In other words, it requires macrospheric immune structures capable of infinite exposition, "global immunizations: co-immunism," a planetary co-immunity wherein the only strangers that are not welcome are those that exploit strangers.[60]

> The history of the own that is grasped on too small a scale and the foreign that is treated too badly reaches an end at the moment when a global co-immunity structure is born, with a respectful inclusion of individual cultures, particular interests and local solidarities. This structure would take on planetary dimensions at the moment when the earth, spanned by networks and built over by foams, was conceived as the own, and the previously dominant exploitative excess as the foreign.[61]

Of course, the task of determining who is exploitative and who is not is full of profound ambiguity and undecidability. It is not as easy as saying yes to community gardens and no to rapacious consumerism. It is not as easy as pronouncing from on high that the Sierra Club stays and Monsanto goes. Building a planetary co-immunity takes practice. The Anthropocene is an era of "anthropotechnics," an era of exercises for humans to learn to coexist with the denizens of a single Earth, exercises in the sense of "monastic rules" or "universal co-operative asceticisms" that are practiced daily to cultivate "the good habits of shared survival."[62] General Immunology thus coincides with a theory of the varieties of anthropotechnic praxis, "General Ascetology," which provides an understanding of the exercises needed for becoming intimately vulnerable at microspheric and macrospheric scales.[63] This is the heart of the Anthropocene, pulsing with microspheric and macrospheric practices of cordial intersubjectivity, anthropotechnic praxes of co-immunity.

A planetary co-immunism is not about choosing between liberal, progressive, social-democratic, or conservative parties. It is about training together to make concrete the universality of intimacy. Such training requires some of the categories celebrated by the Right ("being immune, self-preferential, exclusive, selective, asymmetrical, protectionist") and the Left ("preferential towards the foreign, inclusive, unselective, symmetrical, duty-free").[64] In other words, instead of siding with the Left or the Right, anthropocenic macrospherology demands the development of exercises that facilitate intimately vulnerable relations with the outside and the stranger, exercises that face the aporia of coexistence and facilitate "the implantation of the impossible in the real," the transplant of the outside into the most interior inside.[65] Instead of regular people on the Left and the Right, what is called for is "greatness of soul" (Aristotle), the cultivation of excellence, exceeding the limits of ordinary partisan politics.[66] What is called for is a culture of creatives, developing microspheric and macrospheric exercises that transform the immune systems offered by philosophies and world religions, opening them up to the struggles of planetary coexistence. "Global immunitary reason" is on the horizon.[67]

Global immunitary reason is about struggling for coexistence by remaining exposed, by becoming vulnerable in multipolar intimacies. Composing a global co-immunity entails the radical openness of being-with. I conclude this chapter with a tangible example of a practice of being-with that is also a practice of global co-immunity. The example comes from Donna Haraway, who contrasts sharply with Sloterdijk in some ways, holding very different subject positions and different theoretical orientations (e.g., Sloterdijk has little to do with feminist science studies), but she shares with Sloterdijk as well as Nancy an extended understanding of Heidegger's lonely conception of the openness or world-forming character of human existence.

Haraway differentiates her sense of "the open" from that of Heidegger, who seems to be rather poor in world himself if he is so benighted as to think that animal worlds are inherently poor or lacking in their exposure to beings.[68] Her critique would also apply to Sloterdijk and Nancy as well, although to a lesser degree. Sloterdijk and Nancy extend Heidegger's open, but their attention to the specificities of nonhumans is still rather paltry compared with Haraway's inquiries into nonhuman modes of existence, including thorough analyses of primates, dogs, and genetically modified mice.

Heidegger deploys the term "open" (*das Offen*) to convey the "clearing" that makes room for the disclosure of beings, a disclosure that occurs at the rift between earthly concealment and the unconcealing horizon of a world: "To the open belongs a world and the earth."[69] Heidegger's poetic saying of the open follows from his interpretation of phenomenology in *Being and Time,* where he defines phenomenology in terms of the derivation of the Greek word for "phenomenon" (*phainomenon*) from *phainesthai,* a verb signifying "to show itself," such that "phenomenon" means "that which shows itself, the manifest [das, was sich zeigt, das Sichzeigende, das Offenbare]."[70] The manifest (*das Offenbare*) is what is revealed, what shows itself in the world, in the open (*das Offen*). Contra Heidegger, Haraway's dog has a share in the open that is not poor. It is a share in worlding that is differently open than a human, but that difference could not be described as "more" or "less" open without presupposing one being as the measure of the other, in the way that Heidegger presupposes the human world as the measure for nonhuman worlds and thus fails to understand nonhuman beings on their own terms. Haraway is not speculating. Her articulation of the world-forming capacities of dogs has come about through practice, specifically the practice of companioning with dogs. In a thinking that is a thanking, Haraway takes dogs to heart. It is Haraway's companion species relationship that provides an example of how remaining exposed in multipolar intimacy can facilitate global co-immunity. In the beginning of *When Species Meet,* declaring her intention to nurture "a more just and peaceful other-globalization" ("*autre-mondialisation*") that counters "militarized neoliberal models of world building" without resorting to reactionary "anti-globalization" ("*alter-globalisation*"), Haraway enumerates two guiding questions for her book: "(1) Whom and what do I touch when I touch my dog? and (2) How is 'becoming with' a practice of becoming worldly?"[71]

When she touches her dog, she is touching a companion, cultivating multispecies kinship, multipolar intimacy. Touching is complex, involving connection *and* separation, making contact while tactfully leaving intact. It is not only a matter of petting a pet. The term "pet" is a euphemism for a rather objectifying relationship, reducing the intact otherness of the companion to an object of stroking, patting, rubbing, and fondling. The word is a nominalization of a verb of which the animal is always the object, never the subject. The term can be used affectionately and respectfully, but it is important to demarcate its limitations for thinking through the complex contact between companions. Companioning with a companion animal, Haraway recalls the Catholic sacramental practice of communing with God by eating, sharing in a meal with bread (*cum panis*), where the material Flesh of the bread is simultaneously the semiotic Word (*logos*) of God. The sacramental consciousness of that "hearty meal" is still with Haraway, as she takes to heart the material-semiotic threads woven into coexistence.[72] She takes to heart the task of companioning and communing across the fabricated divides between natures and cultures, humans and nonhumans, organic and technological, facts and fictions.

The way touch takes the other to heart while letting the other remain intact is evident in the Christian biblical scene where the recently arisen Jesus tells Mary Magdalene, "Do not touch me" (John 20:17). For Nancy, this scene makes a sensitive point about touch in general. To truly make contact with another being, to touch with tact, one must paradoxically leave untouched everything one touches. One must open up to the "gap intrinsic to touch," to an "insurmountable edge-to-edge" relationship that unites while differentiating.[73] Touch is like love in that regard.

> Love and truth touch by pushing away: they force the retreat of those whom they reach, for their very onset reveals, in the touch itself, that they are out of reach. It is in being unattainable that they touch us, even seize us. What they draw near to us is their distance: they make us sense it [*sentir*], and this sensing [*ce sentiment*] is their very sense. It is the sense of touch that commands not to touch. . . . You hold nothing; you are unable to hold or retain anything, and that is precisely what you must love and know.[74]

Touching companion species means touching significant otherness, which is stranger than anything that can be grasped.

Haraway is interested not in touching the otherness of an unbelievably omnipotent *God* but, anagrammatically, she is more interested in touching her *dogs*, "Cayenne and Roland," whom she calls "the dogs of my heart?"[75] In particular, it is Cayenne with whom Haraway companions most intimately: "a dog, the dog of my heart, Cayenne, a Klingon warrior princess who was bred to be a working Australian shepherd."[76] This is not to say that they are working as opposed to playing. They are often training for agility competitions, sporting events in which dogs and their handlers move together through a variety of situations and negotiate different kinds of obstacles. Roland is a Chow mutt, but Haraway is more of a working dog at heart, like a border collie, and thus more attuned to the rhythms of Cayenne: "working with Cayenne is so different and partly because I am a border collie at heart myself, but I am learning to pay better attention to Roland's rhythms."[77] Becoming vulnerable to the intimate alterities of multispecies kinship, Haraway describes "companion-species knots in my heart and mind."[78] Her dogs compose her subjectivity, recapitulating on an individual level what took place at the species level: dogs became dogs and humans became humans together in multispecies knots, or in the Aboriginal terms of Rose's multispecies ethnography, "dingo makes us human."[79] Companion species relationships compose intercordial subjectivity. Furthermore, that intercordial subjectivity is multiplied and distributed across dogland—an informal name for the international community of dog lovers, in other words, dog people.[80] Moreover, this heart-to-heart intimacy is something Haraway saw in her mother, whose own sacramental consciousness was evident in the ways she would share suffering, "taking people's needs and pain to her own heart."[81] *Hoc est corpus meum; hoc est cor meum.*

Unmaking and remaking herself and her dogs in companion species relationships, her heart changes, her *cor,* which is closer to her than she is to herself. Her dogs are in her heart, a heart reshaped by attuning to microspheric joy, needs, and pain, a heart that had to grow to accommodate all of the macrospheric complexities of dogland. Surely a growing heart must hurt. Taking to heart more finitude, suffering, and death is painful and depressing. "Intimacy," as Morton observes, "is never so obvious as when we're depressed. . . . To be intimate with the strange stranger is to be in various kinds of pain."[82] The story of that pain is the story of intimate alterities, becoming with strangers, "being-with-being-itself."[83] It is the story of companioning across whatever boundaries might seem to separate human and nonhuman, natural and cultural, material and semiotic. By refusing to undo the story of intimate alterities, by staying with the troubling making and unmaking of companioning bodies, practices of becoming-with are practices of becoming worldly, building a world not according to the story of anatomical individuals sliced up by scholars but according to the story of intimate strangers. "The documenting practices of scholarship slice the heart, but they cannot undo the story."[84]

As a practice of becoming worldly—"cosmopolitics"—becoming-with is "a practice for going on, for remaining exposed to consequences, for entangling materially with as many of the messy players as possible . . . with complexity, care, and curiosity."[85] That complexity entails recognition of complicity. For instance, companioning with a dog involves complicity in unhealthy breeding practices, the commodification of dogs, and the technoscientific instrumentalization of dogs. Haraway's response to that complicity is not critique and cynicism but care and curiosity while remaining exposed to consequences, sharing the suffering that ensues. The world in this context is not a background, a physical or metaphysical container. Indeed, it is not given at all. It is a world in the making. It is a ball of care, a chaotic complex of companions, a messy mesh of strange strangers. Remaining exposed, in the open, practices of becoming-with are practices of becoming worldly.[86] Put quite simply, they are practices of "learning to play with strangers."[87]

If "our hearts are restless" (*inquietum est cor nostrum*), good.[88] That restless feeling is something like indigestion from companioning with messmates. It hurts, but it is not something to avoid. That restlessness is how our hearts attune to one another, attuning to the absurd tunes of strange strangers. That restlessness is how it feels to be in the swarm of coexistence, a restlessness growing with the urgency of ecological emergency. Will our hearts be able to grow accordingly, like miraculously happened with the infamous Grinch? Remaining exposed, there will surely be much heartache and heartbreak in any affirmation of the restless swarm of coexistence during a time of ecological emergency. Vulnerable to the unbearable weight of intimate alterities, it hurts so much to be. Reassuring us that our hearts can bear this unbearable restlessness, the ecological thinker and activist Joanna Macy articulates a meditative practice of "taking heart" informed by her Buddhist practices: "Should you feel an ache in the chest, a pressure in the rib cage, as if the heart would break, that is all right . . . they say the heart that breaks open can hold the whole universe. Your heart is that large. Trust it. Keep breathing."[89]

The suffering, syncopated hearts of the Anthropocene require cardiopulmonary rehabilitation: keep breathing. The practice of breathing is a way of training to remain exposed, open, with . . . Circulating, breath transforms, in becomes out, and out becomes in, and you find yourself inside out—inhabiting the aporia of coexistence, finding yourself losing yourself, finding others and losing them, becoming connected and becoming stranger.

NOTES

1. Martin Heidegger, *What Is Called Thinking?*, trans. J. Glenn Gray (New York: Perennial, 1976), 144. For an overview of the topic of thanking and taking to heart in Heidegger's thinking, see Robert E. Wood, "The Heart in Heidegger's Thought," *Continental Philosophy Review* 48.4 (2015).

2. Martin Heidegger, *The Fundamental Concepts of Metaphysics: World, Finitude, Solitude*, trans. William McNeill and Nicholas Walker (Bloomington: Indiana University Press, 1995), 196.

3. Levinas, *Totality and Infinity*, 74.

4. Morton, *The Ecological Thought*, 51.

5. Thomas Carl Wall, *Radical Passivity: Levinas, Blanchot, and Agamben* (Albany: SUNY Press, 1999). Morton, *The Ecological Thought*, 94.

6. Nancy, *Being Singular Plural*, 38.

7. Rainer Maria Rilke, "A Walk," in *The Winged Energy of Delight: Selected Translations*, ed. and trans. Robert Bly (New York: Perennial, 2005), 176.

8. Augustine, *The Confessions of Saint Augustine*, trans. John K. Ryan (New York: Doubleday, 1960), 43.

9. Rose, *Wild Dog Dreaming*, 15, 21, 112.

10. Susanne Niemeier, "To Be in Control: Kind-Hearted and Cool-Headed: The Head-Heart Dichotomy in English," in *Culture, Body, and Language: Conceptualizations*

of Internal Body Organs Across Cultures and Languages, eds. Farzad Sharifian et al. (Berlin: Mouton de Gruyter, 2008), 367.

11. Sloterdijk, *Bubbles*, 102.
12. Ibid.
13. Ibid., 103.
14. Ibid., 109.
15. Ibid., 110.
16. Ibid., 540.
17. Ibid., 541.
18. Ibid., 335.
19. Heidegger, *Being and Time*, 140.
20. Ibid., 146.
21. Sloterdijk, *Bubbles*, 9, 13.
22. Ibid., 11.
23. Ibid., 340.
24. Ibid., 46.
25. Ibid., 78.
26. Ibid., 28.
27. Ibid., 46.
28. Sloterdijk, *You Must Change Your Life*, 451.
29. Sloterdijk, *Bubbles*, 73.
30. Ibid., 127.
31. Ibid., 129.
32. Ibid., 130.
33. Ibid.
34. Ibid., 137.
35. For multiple accounts of Nancy's transplant, including Nancy's own account alongside other analyses of its biographical and philosophical implications, see the special issue of *The New Centennial Review* edited by Scott Michaelsen and David E. Johnson, *At the Heart: Of Jean-Luc Nancy*.
36. Jean-Luc Nancy, *Corpus*, trans. Richard A. Rand (New York: Fordham University Press, 2008), 162.
37. Ibid., 167.
38. Ibid., 164, 168.
39. Sloterdijk, *Bubbles*, 100.
40. Nancy, *Corpus*, 169.
41. Ibid., 170.
42. Ibid.
43. Ibid.
44. Nancy, "The real outside is 'at the heart' of the inside: Interview."
45. Tom Cohen, Claire Colebrook, J. Hillis Miller, *Twilight of the Anthropocene Idols* (London: Open Humanities Press, 2016).
46. Bruno Latour, "The Anthropocene and the Destruction of the Image of the Globe," Fourth Lecture in the series *Facing Gaia: A New Enquiry into Natural Religion*, February 25, 2013.
47. Ibid., 88.
48. Sloterdijk, *Bubbles*, 617.
49. Ibid., 23.
50. Ibid., 24.
51. Peter Sloterdijk, *In the World Interior of Capital: For a Philosophical Theory of Globalization*, trans. Wieland Hoban (Malden: Polity Press, 2013), 161.
52. Sloterdijk, *Bubbles*, 23.
53. Ibid., 25.
54. Ibid., 24.
55. Ibid., 71.

56. Sloterdijk, *In the World Interior of Capital*, 196–98. Adapting the term *Weltinnenraum* from Rilke, for whom it connotes a sort of cosmic intimacy, Sloterdijk applies it to his analysis of the artificial hothouse built by global capitalism.
57. Sloterdijk, *Bubbles*, 76.
58. Ibid., 23, 69.
59. Ibid., 616.
60. Sloterdijk, *You Must Change Your Life*, 452. Co-immunism has some affinity with the reworked sense of communism developed in Nancy's notion of an "inoperative community." The withness—the "com-" (*cum*)—of the community is not a mere togetherness or simple unity, but is rather "shattered love"—the vulnerable pulsing of cordial intersubjectivity. Jean-Luc Nancy, *The Inoperative Community*, ed. Peter Connor (Minneapolis: University of Minnesota Press, 1991), 82.
61. Sloterdijk, *You Must Change Your Life*, 451.
62. Ibid., 452.
63. Ibid., 110.
64. Sloterdijk, *In the World Interior of Capital*, 263.
65. Ibid., 264.
66. Ibid.
67. Sloterdijk, *You Must Change Your Life*, 451.
68. Donna J. Haraway, *When Species Meet* (Minneapolis: University of Minnesota Press, 2008), 221, 334, 367.
69. Martin Heidegger, "The Origin of the Work of Art," in *Off the Beaten Track*, trans. Julian Young and Kenneth Haynes (Cambridge: Cambridge University Press, 2002), 30–31.
70. Heidegger, *Being and Time*, 51.
71. Haraway, *When Species Meet*, 3.
72. Ibid., 18.
73. Jean-Luc Nancy, *Noli Me Tangere: On the Raising of the Body*, trans. Sarah Clift, Pascale-Anne Brault, and Michael Naas (New York: Fordham University Press, 2008), 13.
74. Ibid., 37. On the profound significance of touching in Nancy's philosophy, see Jacque Derrida, *On Touching—Jean-Luc Nancy*, trans. Christine Irizarry (Stanford: Stanford University Press, 2005).
75. Haraway, *When Species Meet*, x.
76. Ibid., 175.
77. Ibid., 185.
78. Ibid., 97.
79. Deborah Bird Rose, *Dingo Makes Us Human: Life and Land in Australian Aboriginal Culture* (Cambridge: Cambridge University Press, 2000). For an overview of the varieties of coevolutionary relationships between humans and dogs, see Raymond Coppinger and Lorna Coppinger, *Dogs: A New Understanding of Canine Origin, Behavior and Evolution* (Chicago: University of Chicago Press, 2001).
80. Haraway mentions "encounters in dogland, with people and dogs, that have reshaped my heart, mind, and writing." Ibid., 182.
81. Ibid., 172. This passage goes on to mention that Haraway's mother died of a heart attack. It serves as a poignant reminder that this hearty and heartfelt language is not incidental. The vital vulnerability it conveys is literal and not merely literary.
82. Morton, *The Ecological Thought*, 94.
83. Nancy, *Being Singular Plural*, 38.
84. Haraway, *When Species Meet*, 178.
85. Ibid., 106.
86. The contrast between Haraway's sense of "world" and the sense of "world" in other thinkers in proximity to coexistentialism is discussed in chapter 9 ("Becoming Worldly").
87. Haraway, *When Species Meet*, 243.
88. Augustine, *The Confessions of Saint Augustine*, 43.

89. Joanna Macy, *World as Lover, World as Self: Courage for Global Justice and Ecological Renewal* (Berkeley: Parallax Press, 2007), 129.

FIVE

Roundness

What happens when a person becomes well rounded? It is not that they become physically round, although well-roundedness does not necessarily preclude a rotund figure. Typically, a well-rounded person is one who has diverse interests and engages in various aspects of existence with a relatively even distribution. If you focus entirely on one practice, you are not rounded at all. You become excessively acute, thin, or narrow-minded. Imagine someone whose time is occupied exclusively by the study of philosophical texts, with no room left over for music, cooking, sports, friendship, community, or time to companion with a cat. Of course, it is almost impossible to occupy one's time exclusively with one activity. Other activities are bound to enter one's life and add some degree of roundness. Even the most narrow-minded person admits of some degree of roundness. Even if you have no interest in cooking and have no aesthetic appreciation for food, and the only thing you eat is a liquid meal-substitute that you buy in bulk, you are still undertaking a culinary and dietary practice, however meager. Even if your taste in music does not exceed the limits of a handful of pre-programmed Internet radio stations, music probably still plays some part in your life, thus adding some roundness to your existence, however slight. The task of becoming a *well*-rounded person is not just about undertaking a variety of activities and having diverse interests but on maintaining an even or balanced distribution among those activities and interests, so that no domain is out of shape, whether too bulbous, thin, or amorphous.

A well-rounded person is a well-developed person in the sense of someone fully formed, fully in shape. Getting in shape requires *askesis*—"training," "exercise" (see chapter 10). Despite the fact that there are fitness instructors, gurus, coaches, teachers, priests, and many other authorities who would claim that they know how to get a human being in shape, the details of such *askesis* are not self-evident. The way you train to become well-rounded shapes and is shaped by your sense of the training being that has been called "human" (chapter 7), including the sexual differences composing the life in which that being has a share (chapter 8), and the emergency facing the world in which that being dwells (chapter 9). In other words, the way you become well rounded involves a sense of who you are—a human, a sexual being—and what you are doing here—in the world. Furthermore, the way you become well rounded depends upon the character of the roundness that is presupposed by well-roundedness.

Well-roundedness is touted as a value in educational institutions. For instance, the core curriculum in schools supposedly ensures that students do not focus exclusively on one area of study. In Jesuit universities, education is directed toward a care for the whole person (*cura personalis*), including intellectual study and mental health along with physical health and spiritual fulfillment. However, core curricula and holistic pedagogy are regulative ideals whose manifestations still include a lot of overspecialized and narrow-minded people. To some extent, this overspecialization is symptomatic of human existence within individualistic milieus. Acutely delimited individuals are forced to narrowly focus their education and livelihood to such an extent that they lose any sense of roundness and, indeed, lose themselves. Think about the meal-replacement shakes that have been developed for computer coders who focus so exclusively on coding that they suffer from undernourishment due to their lack of interest in food or eating. Nobody is really supposed to be well rounded. They are supposed to have enough roundness so that they do not lose their ability to continue participating in their overspecialized lifestyle, to continue coding to make new websites, apps, and software. Current values of well-roundedness are not really round. They are scaffolding for hyperactive individualism.

Coexistentialism provides a more well-rounded sense of roundness. More than an arrangement of diverse interests, more than just a metaphor, roundness has something to do with the shape of intimate relationships, the intimate curvature of cohabitation, the space of the co- in coexistence, the space of being-with. Whereas the previous chapter takes to heart the intimate vulnerability that comes with the exposure of being-with, this chapter outlines the spatiality of being-with, particularly by following the curves of coexistential spatiality from Heidegger to Sloterdijk, tracing the history of roundness from ancient spheres to the global climate emergency, and considering how to make place for planetary coexistence.

COEXISTENTIAL SPATIALITY

Although *Being and Time* made Heidegger famous for thinking through time, Heidegger makes important contributions to thinking in space in that book and throughout his writings.[1] For Heidegger, the being that calls itself "human" is carefully attuned, inhabiting a world with other beings. That attunement is open to things. It takes them to heart. Being-with means bringing near. "*In Dasein there lies an essential tendency toward closeness*," that is, a tendency toward "*making room*," giving things space to be somewhere.[2] This closeness opens the horizon in which things have room to show themselves, appearing as near or distant, here or there. This is the "inhood" (*Inheit*) that characterizes human existence as being-in-the-world.[3] Opening, making room, the being-in of being-there constitutes the spatiality of existence. Determinations of space as a geometrical form or a physical container are abstractions derived from this existential closeness. Even when philosophers attempt to define space in terms of abstractions, their existential closeness to actually existing things irrepressibly shows itself. What a philosopher attempts to present as a theory of space becomes a theory of some specific thing with which the philosopher is familiar: a house, a womb, folding fabric, or in the case of Descartes, a sponge.

In his thinking on space as extension, Descartes argues that there is a generic, abstract unity to space, indifferent to the changes of bodies in space. When bodies in space change size and shape, the extension of space itself remains unchanged. The thing changed, not space. How does Descartes make this argument? His reasoning requires a sponge: "when we see a sponge full of water or another liquid, we do not think that, in terms of its own individual parts, it has more extension than when it is compressed and dry; but only that its pores are more open, and that its parts are therefore spread over more space."[4] Reflecting on this, Gaston Bachelard observes that "examples can be found where very great minds are stuck, so to speak, in primary images. The metaphysics of space in Descartes is the metaphysics of the sponge."[5] As a primary image, the sponge is a phenomenon showing something with which Descartes is familiar. His attempt to define a supposedly abstract space could not escape the orbit of existential closeness. The point here is not just that philosophers use metaphors to make arguments. It is less about uncovering the metaphors hidden in philosophy than attuning to the philosophy resonating in those metaphors. There is an "intellectualism of metaphor," such that metaphors are "not mere metaphors."[6] This means that an understanding of elements requires more than chemistry and physics. It requires an understanding of the poetics of the elements, including the poetics of earth, air, fire, and water. Each element "is profoundly and materially a *system of poetic fidelity*."[7] This means that space too is poetic, metaphorical, aesthetic. What Bachelard finds in his phenomenological inquiry into spatial poetics is that space is not homogeneous. It is intimately contoured, and arguments to the contrary evaporate in the warmth of existential closeness. Space is more like the familiarity Descartes has with sponges than it is like a generic, abstract unity of extension.

While Bachelard resonates more with Heidegger's existential spatiality than the abstract space of Descartes, he also attempts to adhere more attentively than Heidegger to the intimate contours of things themselves. There are some limitations in the intimacy of Heidegger's *Inheit*. Heidegger's articulations of the intimacy of inhood are lacking in specificity and diversity, from his early works that speak of existence with the highly technical style of hermeneutic phenomenology to his later works, where his poetic sayings gather together a narrow selection of provincial places, like bridges, untrodden paths, and forest clearings. Bachelard's analysis of spatial imagination avoids Heidegger's abstractness and narrowness. In *The Poetics of Space*, Bachelard considers a great variety of spaces, including shells, nests, drawers, cellars, basements, and corners as well as the "inner immensity" or "intimate immensity" of vast horizons and cosmic expanses.[8] Bachelard's sense of intimate space opens existential spatiality to coexistential spatiality. *In each being there lies an essential tendency toward closeness.*

Each being is the center around which a horizon opens and makes room for others to become close. "In this coexistentialism every object invested with intimate space becomes the center of all space. For each object, distance is the present, the horizon exists as much as the center."[9] What Heidegger said of *Dasein* is true of anything in existence, not just humans. If only Heidegger would have used the word "*Dasein*" in its usual German sense as "existence," as "the being" (*das Dasein*), not the "being-there" (*Da-sein*) of human being. Along those lines, Jaspers provides an instructive saying, one that proves to be crucial for Bachelard and even more so for Sloterdijk. Using the word "*Dasein*" in its usual sense ("existence"), Jaspers says the following. "Every being [*Dasein*] seems in itself round, each closed as this in its world, but it is unclosed, because it is possible only by another and with respect to another existence, and because it is in transformation."[10] Each being is its own world, wrapping around itself, bringing close whatever shows up in its horizon. Bringing close, that roundness is closed (*geschlossen*), irreducible to the world of another, and yet that roundness is also un-closed (*un-geschlossen*), as a being is open to change through relations with others and relations with itself.

Although this roundness is a philosophical concept and not merely a metaphor, it is important not to take it too literally. A response to a theory of spatiality as roundness could be to ask where all of the straight lines and angles are in such a theory. First of all, straight lines are technically a type of curve, having a null curvature. Secondly, and more to the point, roundness is not mathematically literal. The curvature is about intimacy, bringing close as including—closing in, wrapping around. The intimacy of roundness can exist with right angles and straight lines, like the intimate space a child might seek out in the corner of a cramped attic, exploring the profound absurdity and uselessness of boredom.[11] The whole point of roundness is to attune to coexistential spatiality and avoid reducing space to geometric literalism or mathematicism. Accordingly, Bachelard notes that the main difficulty he had in articulating a phenomenology of roundness "was to avoid all geometrical evidence. In other words, I had to start with a sort of intimacy of roundness."[12]

Starting with intimacy means attending to images of roundness while holding in suspense the thoughts, feelings, and actions with which one habitually engages in spatial relationships. For Bachelard, this requires radicalizing the roundness of Jaspers so that each being does not *seem* round but *is* round: "*das Dasein ist rund*, being is round."[13] Bachelard follows Jaspers in avoiding Heidegger's anthropocentrically specialized *Dasein*. To say that being is round means any and every being, including a bird, a walnut tree, and a closet as much as a human. "The evening lamp on the family table is also the center of a world."[14] Bachelard modifies what Jaspers says because the idea that every being seems (*scheint*) round puts roundness on the side of appearing and not being. For Bachelard, intimacy cannot be understood in terms of the metaphysical doublet of being and appearing. That doublet cannot account for intimacy; rather, intimacy accounts for that doublet.

Whether Bachelard's rephrasing really avoids the being/appearing doublet seems doubtful, since it appears to be trading seeming for being. Perhaps a more Heideggerian reading of Jaspers' phrase would be appropriate, as it gathers multiple senses of *Schein*.

> 1) Schein as radiance and glow; 2) Schein and Scheinen as appearing, as coming to light; 3) Schein as mere appearance or semblance ‹Anschein›. But at the same time it becomes clear that the second variety of "Scheinen," appearing in the sense of showing itself, pertains both to Schein as radiance and to Schein as semblance, and not as a fortuitous attribute but as the ground of their possibility. The essence of appearance ‹Schein› lies in its appearing ‹Erscheinen›.[15]

The sun is one of Heidegger's examples here. It shines, showing itself, radiating light and warmth, and seemingly (i.e., not really) moving around Earth. Glowing with a Heideggerian shine, Jaspers' phrase ("Jedes Dasein scheint in sich rund") would mean that "every being shows itself in itself round." Showing itself, a being *is* in itself round, and that roundness shines with aesthetic radiance and the ambiguities of semblance.

In any case, the point here is less about the terms used than about the intimacy with which the rift between what something is in itself and what it appears to be opens up. The gap between semblance and substance is characteristic of the closeness of being-in. When I see a rope and mistake it for a snake, the gap between its serpentine appearance and its rope reality opens insofar as I am already close with the rope, exposed to the disclosure of its roundness. To say every being is round is to say that every being has a tendency toward closeness, which brings things near in the undecidable ambiguity of their appearing reality, which is also the ambiguity of my appearing reality. It is the profound ambiguity of the relational contact and intangible alterity of a being.

Although he does not use the image of roundness, Morton makes a similar point about space in his object-oriented ontology. Space is not homogeneous. It is not extension, geometrical form, or mathematical abstraction. Space emerges from things themselves as they relate to one another, getting close with one another. Space is the poetics of the intimate intertwining of objects, the metaphors whereby they interpret and translate each other.

> Thus to be located "in" space or "in" time is already to have been caught in a web of relations. It is not that objects primordially "occupy" some existing region of spacetime, but that they are caught in the fields of, and otherwise "spaced" and "timed" by other entities.[16]

Space is emitted from objects. It is exposure of intimate alterities enmeshing one another. Morton points out that this conception of space "is congruent with the last century of physics. For Einstein, space and time are also emergent properties of objects: objects don't float in a neutral void but emanate waves and ripples of spacetime."[17] To be in space, at the inn of space, is not to be in a box or on a background. It is to be in close relations with others, intimate alterities. This means that spatial relationships are nonlocal. The here and there of me and a glass of wine are already intertwined by the time I decide to reach for it and take another drink. It is only because of our intimate proximity that I can overcome the distance between us by reaching for it and successfully grabbing it, or unsuccessfully grabbing it, overreaching and knocking the glass over.

For Heidegger, spatiality is reserved for relationships that a human might have with a sponge but not the relationship that a tree might have to a bird or a sock might have to a drawer. Outside of the horizon of human existence, there is no room for beings. Jaspers, Bachelard, and Morton facilitate a more well-rounded redistribution of existential spatiality, recognizing that all beings have a tendency toward closeness. In the context of coexistential spatiality, the emergence of colonialism, globalization, and global warming are all part of a single crisis of roundness, a crisis of intimacy pervading the spaces of the Anthropocene, including spaces of geoinformatics, landscape architecture, urban planning, sacred sites, sustainable design, migration (human and nonhuman), habitat destruction, ecological restoration, wilderness preservation, websites, and "safe spaces" for the social and political participation of marginalized groups. It is "a general space crisis."[18] Humans must learn to exist in a single orb, Earth—the last orb around after the destruction of the spheres of God and nature. Expanding on Heidegger's notion of being-in and Bachelard's notion of roundness even further, Sloterdijk gives such a history of modernity as a history of roundness, a history situated in his account of the "multi-significant universes of existential spaciousness" he calls "spheres."[19]

Things do not merely *seem* round for Sloterdijk. His theory of spheres attempts, through "a slowed-down circumspection," to discern "something that is the most real, yet also the most elusive and least tangible of things."[20] Sphere construction is that whereby existence is also coexistence, inhabitation is cohabitation, being oneself is being outside of oneself in a mesh with others. "We are in an outside that carries inner worlds."[21] For humans, a "sphere is the interior, disclosed, shared realm inhabited by humans—in so far as they succeed in becoming humans. . . . Spheres are immune-systemically effective space creations for ecstatic beings that are operated upon by the outside."[22] To tell the story of planetary coexistence, Sloterdijk tells a story of spheres.

A GLOBAL NARRATIVE

Does anyone want to hear a grand narrative? On one hand, there is the most common answer, which asserts that grand narratives have been relegated to the realm of the incredulous, having suffered numerous blows by philosophers with hammers of critique, deconstruction, and deterritorialization. This allergy to grand narratives is exemplified in Jean-François Lyotard's response to the postmodern condition, which he defines as "incredulity toward metanarratives," preferring a "little narrative" (*petit récit*) to the seemingly totalizing aspirations of grand narratives.[23] On the other hand, the death of metanarratives has been greatly exaggerated.

Even though many theorists talk about living in a cynical era characterized by widespread disbelief in big pictures and big stories, people are still practicing metanarratives, especially the grand narrative of global capitalism, a narrative of the emergence and triumph of the consumer subject. If Lyotard represented the previous position, Slavoj Žižek represents this position.

> So, when we are bombarded by claims that in our post-ideological cynical era nobody believes in the proclaimed ideals . . . one should always counter such claims with the question: OK, but *where is the fetish which enables you to (pretend to) accept reality "the way it is"*? . . . It enables you to fully participate in the frantic pace of the capitalist game while sustaining the perception that you are not really in it.[24]

Furthermore, metanarratives survive not only in everyday practices but also in theoretical articulation. Philosophers never actually stopped proposing grand narratives. Exemplary in this respect would be the comprehensive narrative of epistemic change that Foucault articulates in his archaeology of knowledge, or more recently, Ed Casey's philosophical history of the emergence, loss, and recovery of place in occidental thought, or the *Empire* trilogy by Hardt and Negri, which presents a grand narrative of the repression and liberation of the multitude.[25]

Sloterdijk integrates these various responses to the problem of grand narratives. He is like Lyotard in recognizing widespread cynicism and incredulity toward all sorts of metanarratives and comprehensive systems, but Sloterdijk does not stop there.[26] He is also like Žižek in recognizing that such cynicism is symptomatic of the consumer subjectivity propagated through a neoliberal grand narrative, and he is also like Foucault, Casey, and Hardt and Negri in attempting to articulate a new grand narrative. For Sloterdijk, while the task of constructing a grand narrative is surely provocative, giving up on such a task would be "an act of intellectual defeatism," not to mention that it does not face its self-contradiction: the meta-narrative that there are no meta-narratives; the absolute position that there are no absolutes.[27]

In what follows, I follow Sloterdijk in the audacious task of proposing a new grand narrative. This metanarrative, unlike those which garner the incredulity of Lyotard, is not a narrative of the progress of knowledge or a narrative of emancipation. However, even though it posits neither universal progress nor emancipation, it is not just a little narrative (*petit récit*). Its scope is too large and its aim too ambitious. The scope of the new grand narrative is Earth, or more specifically, "the encounter between being and the circle" in globalization, and the aim of this narrative is to contextualize the composition of a global immune system—a place for ecological coexistence on one Earth.[28] The encounter between being and the circle takes place on microspheric scales (bubbles) as well as macrospheric scales (globes). Bubbles are "intimate forms of the rounded being-in-form and the basic molecule of the strong relationship."[29] Bubbles indicate that the human is never alone. There are always at least two. Dipolar intimacy is the minimal unit of human existence. Such microspheric spaces include intrauterine existence, face-to-face relations, magical bonding (hypnosis, transference), genius and angelic relationships, the resonant space of the voice, and heart-to-heart relations (i.e., the intercordial subjectivity explored in the previous chapter, "Remaining Exposed"). Sloterdijk's grand narrative is a narrative of the macrospheric counterparts to bubbles—globes—in which "being-in on a small scale returns as a political and cosmic relationship."[30]

Sloterdijk demarcates three phases of his grand narrative of globalization. First, "cosmic-uranian globalization," which begins with the spheres of classical ontology; second, "terrestrial globalization," which spans the five-century period from the beginning of colonialism (1492) to the end of World War II (1945); and third, "electronic globalization," which refers to the advent of information technologies and new media beginning in the latter half of the twentieth century.[31] The first phase is called "cosmic-uranian" because, during that phase, macrospheric intimacy represented "the totality of what exists in the stimulating image of an all-encompassing sphere," which was called various *cosmos* and *uranos*.[32] The former name connotes the ornamental cosmetics and beautiful order of the universe, while the latter bears titanic connotations of the ethereal limits of this all-encompassing sphere. In the cosmetic cosmos, the sky is the limit. The cosmological meditations of ancient astronomers, mathematicians, and philosophers exemplify the articulation of this macrospheric system. Those ancient cosmologists accomplished a rationalization of space, taking measure of the sky and the motion of things, turning the swarm of coexistence into a coherent whole.

Aristotle is a clear example here. Aristotle notes that two kinds of natural things exist: generated things like plants and animals, and eternal things like heavenly spheres and stars. Rejecting the notions of some of his predecessors and contemporaries who view things as having arisen from "night," chaos, or nonbeing, Aristotle posits the eternality of movement or time. Because it would be contradictory to speak of anything *before* time, time or motion must be continuous, and continuous movement can only occur as circular rotation ("there is no continuous motion other than in place, and among these, other than in a circle").[33] Hence, eternal, heavenly things move in orbital paths. Each sphere moves according to its desire to be like the sphere directly encompassing it, which is to say, the movement of spheres exists through the imitation of outer spheres, and the perfect motion of the outermost sphere is caused by the perfect activity of the unmoved mover, God.

For Aristotle and through the ancient world, being is well rounded. "Classical ontology was a spherology."[34] The motion of things finds its perfect ground and orientation in macrospheric intimacy. The great orb is variously articulated as the oldest, greatest, most beautiful, wisest, strongest, fastest (through omnipresent mind, *nous*), and divine.[35] Some of the good news of this philosophical arrangement is that humans had a safe operating space in which they could negotiate the stress, depression, and constant motion of coexistence. However, this operating space was troubled by an ambiguity. Theorizing about spheres, contemplating the whole, philosophers affirmed the cohabitation of humans within spherical being, yet they simultaneously took up the position of a transcendental observer capable of understanding spheres from the outside. The construction of macrospheric intimacy is also its destruction. Building the macrosphere coincided with fleeing the macrosphere. Philosophers forgot their actual place in the world in order to affirm its intimate order. "What Heidegger would later term forgetfulness of being [*Seinsvergessen-*

heit] already begins with the ancient instructions for viewing the orb blissfully from without," instructions for "lordly observation of the outside that would one day lead to polytechnical dreams of control and the tyranny of knowledge over concretely interpreted life as a whole": cosmic-uranian globalization thus already sets in motion the "betrayal of humanity's existential place."[36]

The rational and abstract viewpoint of cosmic-uranian globalization led to its modern failure. The imperfections of this world cannot be held by philosophical appeals from outside of the world, appeals to the perfection of an ideal or divine world that transcends this world. How much can an otherworldly immune system protect you? What is needed is a planetary immune system. If the intimate space of a globe is going to sufficiently nurture us, it must be the intimate space of this globe, this actual orb, Earth. Accordingly, as Sloterdijk observes, "the Uranian globalization of ancient physics had to change into terrestrial globalization upon its modern failure."[37] Moreover, it was not only failing because the demands of actual existence cannot be sufficiently addressed by otherworldly ideals. It was failing due to the curiosity and care that characterize the ecstatic openness of human existence. By remaining exposed to beings, humans learned different things about them, and some of those things contradicted the orderly spheres of cosmic-uranian globalization. The Copernican turn is a crucial example, as it marks the end of the classical macrospherology. "With the help of its relentlessly probing intelligence, the open animal tore down the roof of its house from the inside. Taking part in modernity means putting evolved immune systems at risk."[38]

After the Copernican turn, "the sky as an immune system was suddenly useless."[39] This is part of what Nietzsche announced in saying "God is dead": "What were we doing when we unchained this earth from its sun? Whither is it moving now? Whither are we moving? . . . Do we not feel the breath of empty space? Has it not become colder? Is not night continually closing in on us?"[40] This coldness is symptomatic of a climate crisis. The warming intimacy of the ancient spheres was gone—the roof torn off, the air-conditioning system broken. Living in modernity means living with the removal of the immune systems provided by traditional cosmological and religious narratives, living on a sphere (Earth) instead of living in one. Peeling away layers of ancient macrospheres, modernity is characterized by the challenge of constructing new ways of maintaining macrospheric warmth. "The peeled human being acts out its epochal psychosis by replying to external cooling with warming technologies and climate policies—or with climate technologies and warming policies."[41] Insofar as the human species exists because of successful constructions of intimate spaces of coexistence, constructions of "closeness hothouses," then a "'climate catastrophe'—the authentic password for our epoch—

already formulates the original risk for humanity," the original vulnerability of being open, exposed, in the world.[42]

Sigmund Freud's assessment is that the Copernican revolution was one of three "great outrages" injuring the "naïve self-love" of humanity in modernity (the other two being Darwin's theory of evolution and Freud's own psychoanalysis).[43] However, as Sloterdijk points out, the stellar coldness of modernity was not initially received as bad news but more like a breath of fresh air, which "had an inspiring effect on the self-confidence of Europeans."[44] With a boost in European self-confidence and a need for new macrospheric constructions, the second phase of globalization begins: terrestrial globalization. This phase is characterized by "the monogeistic faith of the Modern Age."[45] In this faith, Earth is elevated to supreme status, emerging as "the transcendental heavenly body that, as the condition of possibility of observing the other heavenly bodies, outshines all others."[46] Earth is "the transcendental star that comes into play as the locational condition for all self-reflections. . . . As the star on which the theory of stars appeared, the earth shines with self-generated phosphorescence."[47]

The phrase "terrestrial" is accurate insofar as this period of globalization seeks macrospheric intimacy as inhabitants of one Earth. However, the phrase terrestrial is misleading, as this phase of globalization began with "nautical globalization."[48] Seafaring, exploration, navigation, and cartography led the way for trade, colonialism, and Christian missionary activity. The disasters of this period of globalization are countless. The attempt to build closeness hothouses after the collapse of the classical spheres resulted in massively distributed destructions of intimacy. The intimate spaces of Earth began shrinking, such that closeness was not closeness with difference, alterity, or distance but was precisely the eradication of distance and the homogenization of difference. Terrestrial globalization "de-spatializes the real globe, replacing the curved earth with an almost extensionless point, or a network," and "the dignity of distances is negated."[49] Far from producing an intimate space for planetary coexistence, the hothouse of terrestrial globalization only benefits the very few, while the vast multitude of terrestrial beings are excluded. One of the "thermo-political paradoxes" of globalization is that inhabitants of traditional worlds are "exposed to the frosts of freedom" for the sake of finding warmth. They are being told to evacuate their old spheres for the promise of a better life, but the loss of the immune systems of traditional spheres is devastating to all the beings depending on that immunity.[50] Promising planetary intimacy, terrestrial globalization produced "a comfort-animated artificial continent in the ocean of poverty."[51] To put it another way, the globe of terrestrial globalization is not intimately warm. It is on fire, in the words of Blake, "a red round globe hot burning."[52]

Sloterdijk refers to the global hothouse in terms of the metaphor of a "Crystal Palace" invoked by Dostoevsky in his *Notes from Underground*.[53]

The metaphor is based on a large plate-glass and cast-iron structure built for the Great Exhibition of 1851, which was an exhibition of the state of the art of industrial technology. For Dostoevsky's underground man, the Palace exemplifies a desire for permanent pampering that suppresses and excludes the suffering, doubt, impermanence, and uncertainty of existence.[54] The promise of that "the 'Palace of Crystal' will be built" is the promise that "new economic relations will be established, all ready-made and worked out with mathematical exactitude, so that every possible question will vanish in the twinkling of an eye, simply because every possible answer to it will be provided."[55] Dostoevsky sounds like an antiglobalization activist *avant la lettre*. The crystal palace is not an intimate space of planetary coexistence. Sloterdijk calls it "a planetary palace of consumption," in which human existence is reduced to the pursuit of spending power, and freedom means the freedom to pick and choose between different products and brands or, even better, making your own products and brands.[56] Although the "social-communist project" had its own aspirations for constructing a global hothouse, capitalism was much more powerful at capturing the burning desires for a planetary palace. Living in the crystal palace means living in the *Weltinnenraum* ("world interior") of capital.[57]

The triumph of capitalism took place during the third phase of globalization: electronic globalization. This marks the "saturation phase" of the monogeistic turn, which can be defined in terms of three dimensions: morally, it involves the abandonment of unilateral, colonial, and imperial projects; technologically, it involves social media, information technology, and virtual reality; and systemically, it involves the radical reciprocity of "post-historical civilization," in which people experience the feedback of the consequences of their actions relatively immediately (e.g., celebrities and politicians are immediately chastised on social media for inappropriate statements in interviews).[58] "The faith of the seafarers changed into knowledge, and that knowledge became trivial and specialized; the earth-believers of the sixteenth century are now postmodern."[59] This posthistorical, postmodern, postcolonial information age is one in which all human endeavors congeal in a "hyperactive vibrating jelly."[60] The global greenhouse is a gelatinous goop, not an intimate orb. Individuals cannot find intimacy in the vibrating jelly of global capitalism. Whatever bubbles we inhabit vibrate alongside another, frothing and foaming amid the increasing ruination of our shared climate. The capitalist macrosphere is not really a macrosphere as much as it is a foaming mass of microspheres that are fighting for survival in the midst of climate catastrophe. "In foam worlds, the individual bubbles are not absorbed into a single, integrative hyper-orb. . . . What is currently being confusedly proclaimed in all the media as *the* globalization of the world is, in morphological terms, the universalized war of foams."[61]

PLACE FOR THE PLANET

As the functioning of the hothouse of global civilization continues to exclude and oppress human lives while pushing planetary systems beyond stable thresholds, what is called for is more than just a "safe operating space for humanity."[62] What is called for is the composition of intimate ecological spaces of planetary coexistence. How can coexistentialist conceptions of spatiality contribute to the task of facilitating the flourishing coexistence of the inhabitants and habitats of Earth? There are many possibilities, including the possibility of giving up any kind of existentialist nonsense, leaving behind philosophical inquiry and doing something that seems more useful or practical at the grassroots level, like becoming an environmental activist, permaculture gardener, community organizer, or spending quality time with your cat. There is also the possibility of engaging in philosophical inquiry in a way that is itself a form of macrospheric activism, indeed, as environmental philosopher Baird Callicott puts it, "the most radical and effective kind" of activism, "helping to reconfigure the prevailing cultural worldview and thus helping to push general practice in the direction of environmental responsibility."[63]

Pushing for responsibility, philosophy can become a practice of spherological activism in the cultivation of a philosophical attention to place. This attention to place is not an injunction to inhabit a simply fixed location or conceptually fixate on place. Rather, attention to place can open up possibilities for intensifying one's engagements in the vulnerable intimacy of coexistence, like the exercise of attention (*prosoche*) in contemplative spirituality, which in this case could be described in terms of a "contemplative ecology," adapting the practice of *prosoche* to the context of ecological emergency. In his contemplative ecology, Douglas Christie describes the spiritual exercise of *prosoche*, which figured in early Christian monasticism as well as Stoic and Epicurean schools of philosophy. The aim of *prosoche* is the intensification and expansion of consciousness, which is to say, the cultivation of "heightened awareness" that is "extended, potentially, to one's very sense of what it is to be alive in the world."[64] With intense attention, encounters with place can arouse a sense of ethical responsiveness to other beings. You need not look far to begin such a contemplative practice. You can begin with a glance.

Edward Casey, from a phenomenological perspective, elaborates on what takes place in "the moment of the glance," finding that the glance opens up the "first moment of ethical responsiveness," at least with regard to vision.[65] The glance has been neglected by much of Western philosophy and psychology and is seemingly meager in comparison with traditionally discussed types of looking, such as the gaze of a lover or an artist, the scrutiny of scientific inspection, or the contemplative vision of ascetics or mystics. Although it seems meager, the glance is not a matter of trivialities. Lacking the stability and gravity of the gaze or the stare, the glance graces things with "the lightness of Becoming," a lightness that accompanies other psychological functions, such as our cognitive experience of short-term memory or our immediate emotional presence in the world.[66] Like the glance, such cognitive and emotional events take in an immense amount of complex and subtle information within fractions of a second. Casey describes the lightness of the glance as a sort of "egoless ecstasy" wherein "the glance refuses to succumb to the grasping that is so endemic to any settled sense of self."[67] In other words, the moment of a glance effects a release, however slight, and in this release, a glance does not attempt to grasp the world, as ocularcentric philosophies tend to attempt.[68] A glance is more intimate than that. It "*takes in*"—it goes out and "comes (back) in."[69] It takes in a whole situation, a layout of surfaces, a place, a scene of action. Perhaps a glance can even come (back) in with a world, allowing one, as Blake says, "To see a World in a Grain of Sand."[70]

We see some basic characteristics of the glance by considering what happens when a glance takes in another person.[71] First of all, in a glance, you catch a glimpse of the feelings, interest, and mood of the other person. One does not need to gaze or stare to determine that another person is angry. A glance will suffice, which is not to say that your initial determination is always correct. Second, the glance also has a power to discern depth, to take in a sense of something disturbing or stirring beneath the surface. You can see that I am hiding something. Thus, in a glance one can see not only that a person appears to be happy on the surface, but one can also see that this apparent emotion is a cracked surface or a facade that seems disturbed by tensions occurring beneath the surface. Also, a glance can take in factors of a person's identity and history, including economic class, race, gender, age, ethnicity, and again, this is not to say that one's determinations are correct. Finally, a glance can take in another glance. In other words, a glance can participate in an exchange of glances, with each glance taking in the other. In a glance, I can see you, see you seeing me, see you seeing my seeing, in the infinite exposition of being-with.

It is the paradox of the glance that it is so diminutive and light and yet is able to take in such deep and subtle insight into the other.[72] The lightness of the glance gives witness to the dynamic and compelling radiance of the other, thus preventing the other from being reduced to a set of qualities, processes, or social constructions. Giving witness to the other, the glance gives witness hospitably, taking the other to heart. The glance is welcoming, not grasping. In describing how the moment of a glance is a witnessing and welcoming of the other, Casey is drawing on Levinas, for whom witnessing and welcoming the other are fundamental to the ethical relation—the relation of responsibility, the relation of "the one-for-the-other," which is the "substitution of the same for the other."[73] In substitution for the other, the human subject is responsible for all things. The ego is an egoless ego, an ego for the other, inside out. "The word *I*" according to Levinas "means *here I am*, answering for everything and for everyone."[74] The ego is thus understood as a "*sub-jectum*" in the sense that it is a subject "under the weight of the universe, responsible for everything."[75] The human subject is responsible insofar as it is in proximity to the other, not in the sense of proximity to an object of knowledge or even to a phenomenon, but proximity to the compelling alterity of "the face" of the other.[76]

The face harbors an alterity that is so completely and uniquely other that it cannot be recovered, represented, or even perceived, because perceptions reduce the other to particular qualities and characteristics that are but traces of the sheer existence of the other. However, adopting terminology from Leibniz, the glance is a matter of "petite perception" ("apperception"), which is a complex and subtle perception that witnesses and welcomes finely textured differences rather than rigidifying and totalizing the other.[77] With a glance, the face of the other arouses in the subject the ethical obligation of responsibility, the obligation to substitute oneself for the other, to put oneself in the place of the irreplaceable other. One can explain the face in terms of perception and appearance, but in the lightness of the glance, the face arouses a sense of responsibility that is irreducible to any particular qualities. This raises an important question for Levinasians. If the face cannot be reduced to particular perceivable qualities, does that mean that the face can be glimpsed on a tree, a computer, a river, a virus?

According to Levinas, for the other to be received ethically, that is, in its complete otherness, the other "must be received independently of his qualities."[78] Therefore, the face should not be restricted to human faces to the exclusion of nonhumans, for such a restriction would reduce the face to its qualities rather than witness and welcome its irreducible otherness. Although Levinas suggests that all things bear traces of the face, and that responsibility answers to the otherness of everything, he mostly speaks of the face in anthropocentric terms—the face is the human "face of a neighbor."[79] Such a perspective is problematic for environmental ethics insofar as an ethical response to the environment entails a movement away from anthropocentrism, that is, away from the view that responsibility to nature, life, and the cosmos is or should be peripheral to responsibility between humans. Thus Levinas' account of the face, while deeply insightful in its treatment of the ethical relation between humans, is in some sense too limiting to describe relations of responsibility to ecological or technological systems or to the world as a whole.[80]

A glance takes in more than just the faces of our human neighbors. One can find a world with a quick glance at a grain of sand. One can find a world in a place, or as Casey calls it, the "place-world." In a glance, the constituents of a place-world call for responsibility to its irreducible and irreplaceable otherness. More inclusive than Levinasian faces, place-worlds "are the ultimate source of ethical force in the environmental field."[81] When a thing calls for responsibility, it does so in light of the place in which it is embedded. To understand specifically how taking a glance at a place can call for responsibility, Casey draws on James J. Gibson, whose ecological approach to perception can be understood as an extension of the radical empiricism expressed by William James.[82] Like James, Gibson seeks to describe perception as it flows within the stream of consciousness.[83] Gibson focuses his description on the ways in which animals and humans visually perceive the environment. In any visual perception of the environment, one perceives places: small places and big places, urban, rural, and wild places, places to move to or from, places from which one could fall, or places for sleeping, eating, meeting, or hiding.

A place-world only appears by means of a landscape, or more generally, a place-scape, which could include cityscapes, mountainscapes, airscapes, seascapes, mediascapes, technoscapes, etc. "Layout" and "surface" are two of the main features of these place-scapes. Gibson uses the term "layout" to refer to "the persisting arrangement of surfaces relative to one another and to the ground."[84] Perception of place is thus always a perception of a layout, a perception of an arrangement of surfaces whereby the sensing body constitutes and is constituted by its environmental conditions. Furthermore, the ground relative to which an arrangement of surfaces is laid out is itself a surface. According to Gibson, "The literal basis of the terrestrial environment is the ground," which is an "underlying surface of support" that tends to be relatively flat and "perpendicular to gravity."[85] Surfaces are the surfaces of substances, things such as rocks, soil, mud, trees, buses, deer, and smog. A perceived surface is an interface between a substance and an elemental medium. A glance at a rock sitting in the middle of a river discloses the interface between the substance of the rock and the air above and water around that provide the medium in which the rock appears. A glance at the surface of the river discloses the interface between the fluid substance of the river's water and the medium of air above it and the land at its banks. Moreover, a glance also discloses depth in the rock or the river, but only insofar as the depth appears somewhere, that is, in a layout of surfaces. You can tell if the river is deep by glancing at the river bottom or lack thereof through the surface of the water. You can feel the density of a rock's interior by touching its exterior. The depths appear at the surface. In the fashion of coexistentialism, things wear their hearts on their sleeves.

An important characteristic of the perception of places is that it discloses various possibilities provided by a place; that is, it discloses what Gibson calls "affordances." "Different layouts," according to Gibson, "have different affordances for animals."[86] For example, water affords habitat for fish, some plants, and microbial organisms, whereas water affords humans swimming or wading but not permanent dwelling.[87] For humans and some animals, the perception of water affords washing, cleaning, and bathing. If it appears in a jug or vessel, water can afford spilling or pouring. If it appears with a hydroelectric dam, water can afford distributing electricity. Affordances *appear* insofar as they are perceived at the surface. Accordingly, in the trio of surface, medium, and substance, it is from surfaces and their shapes, colors, textures, and movements that perception takes place in the place-world. "The surface is where most of the action is."[88]

Perhaps every surface of a place harbors the compelling alterity that Levinas glanced in the face of the other human. In a glance, surfaces are already there, in your face. They provoke us to face the compelling radiance of ecological coexistence. As you follow any surface to the limit, more surfaces show up. A glance at a surface takes in the features of that surface while also taking in other surfaces at the limits of this surface, at the "occluding edges"—the edges that simultaneously unite and divide surfaces, connecting them as they keep them separate, revealing them as they hide them.[89] Like the face of a human neighbor, a place or a surface can, in a glance, arouse one's sense of responsibility for the intimate alterities of planetary coexistence.

To arouse responsibility for the planet's places and to the distressing problems appearing across their surfaces, a philosophical approach to environmental activism practices the glance, seeing the world in a grain of sand, carefully attending to the intimate alterities of every face and surface woven into the mesh of ecological coexistence. For Casey, cultivating such a practice of glancing will help us get back into place.

> Getting back into place, the homecoming that matters most, is an ongoing task that calls for continual journeying between and among places. . . . As travelers on such a voyage, we can resume the direction, and regain the depth, of our individual and collective life once again—and know it for the first time.[90]

Along these lines, Casey's contemplative practice of glancing is aligned with place-based ecological movements, such as bioregionalism. Indeed, bioregionalism has much to contribute to the composition of place-based ways of thinking, feeling, and acting: knowing the source of your food and water, caring for neighboring species, understanding terrains and maps, and participating in local decision-making. A problem with some forms of bioregionalism is a localism or antiglobalism that fails to attend to the rest of the planet, focusing on local places (i.e., ecosystems or bioregions) to the exclusion of other locales and the exclusion of larger scales. A notable exception is Gary Snyder, whose bioregionalism is inflected through the large-scale visions of Buddhism, systems theory, and ecology. "To know the spirit of a place is," for Snyder, "to realize that you are a part of a part and that the whole is made of parts, each of which is whole. You start with the part you are whole in."[91]

There are countless ways to cultivate intimate spaces of planetary coexistence, from contemplative glances to ecosocial movements like bioregionalism. In Bauman's viable agnostic method (discussed in chapter 3), a planetary ethic opens out of the vagabondage of a "polyamory of place."[92] Rose's ecological existentialism emerges out of the multispecies kinship groups inhabiting the nourishing terrains of Aboriginal country, not unlike Haraway's companion species inhabiting "naturalcultural contact zones."[93] Constructing global intimacy can happen by doing something on a computer or in a garden, in a city or on a farm, with a pet or with a wild animal, with a nongovernmental organization or with a corporation. The question is whether you are remaining exposed, vulnerable to intimate alterities, attentive to the finely textured differences of things.

The global hothouse is unjust when its construction fails to let the abysmal dust of this plant remain: just dust. Its injustice takes place when the plural singularities of Earth are effaced, suppressed, or excluded, reduced to instruments of consumerism or covered over with abstractions and calculations. That can be rephrased by saying that the global hothouse is unjust insofar as *place* (heterogeneous singularities) is reduced to *space* (homogeneous abstraction). Place is intimacy, while space is sphere pathology. "Place is the weird one; space is the reified box."[94] In terms of Casey's grand narrative, the global hothouse is the latest instantiation of the modern fantasy of space as supreme container, a supremacy which is criticized and countered in twentieth-century concepts of space-time and in postcolonial and ecological efforts to restore places and reaffirm specificity and locality after centuries of homogenization.[95] "Space," as Morton puts it, "has revealed itself as the convenient fiction of white Western imperialist humans," and that "fantasy of empty, smooth" space has ended, collapsing under the unbearable weight of ecological emergency, and "*place* has emerged in its truly monstrous uncanny dimension, which is to say its nonhuman dimension."[96]

While the distinction between the intimacy of roundness and the injustice of sphere pathology is important, it is not always best described in terms of a distinction between place and space, wherein the latter is an abstraction that alienates one from the real, and the former is a web of relations in which one is immersed along with other strange faces and uncanny surfaces. The space/place distinction has become a convention circulating in phenomenology and place studies, and having become a standard academic category the area of this distinction began to look a lot like a homogenized space. Surely homogenization was not the aim in articulating the distinction between abstract homogeneous space and actual heterogeneous places. A disambiguated determination of place in distinction from space is counterproductive to the attention to place that such disambiguation seeks. If you are really attending to the exigencies of place you are not drawing a mark between place and space as much as you are facing the abyss of the question of "where," remaining exposed to the consequences of your contact with intimate alterities across microspheres and macrospheres. The space/place distinction occludes that abyss, sorting out places and spaces, attempting to exorcise the specter of ambiguity haunting a spectrum of places, areas, spaces, sites, location, milieus, regions, positions, and locales and multiple scales, from quantum to cosmic. Yet, the recalcitrant abyss persists, haunting the boundary between space and place. *Where* is that boundary drawn, in a place or a space?

Concerns about place and its violation in displacement, environmental destruction, neoliberal enclosures, or scientific homogenization and objectification are urgently important, so it might seem that this is not a matter that calls for protracted debate and speculation. However, it is precisely their urgency that calls for thinking. The clear distinction between space and place is one of the things getting in the way of restoring and protecting places. There are multiple reasons for embracing space/place ambiguities. First, anything can be either place or space. They are not two different kinds of things or two different areas as much as they are two different ways of attuning to things. The space/place distinction makes it sound like the difference lies more in the things themselves than in the modes of attunement with which others connect with those things. A computer can be a milieu of intimate experience (place) as well as a programmed matrix of abstraction (space). As Levinas noted, the Bible became a "nourishing terrain" (*espace vital*) for Jews (the people of the Book).[97] National parks are not just places of ecological vitality and nature immersion. They are also sites of rational methods of ecosystem management and the economics of tourist-related industries. Second, by standardizing the language of being-in, this distinction is propagating the reification that it was resisting, thinking too abstractly and from the outside about the importance of thinking in a way that is intimately interwoven with place, like ancient cosmologists looking at spheres from the outside while positing our belonging inside. In a relatively cold and certain tone, discourses on space and place distract from the coexistential struggle of sharing intimacy and uncertainty. Where does the space/place boundary become articulated, in a homogeneous space of sclerotic categorical distinctions? Third, the space/place distinction forecloses the undecidable who or what of coexistential encounter, reinscribing the encounter within dualisms of animate and inanimate, abstract and concrete, heterogeneous and homogeneous, natural and cultural, immersion and alienation, and so on.

Finally, consider the sorites paradox, which is associated with vague predicates, such as a "heap." When is a heap a heap? If I put two pebbles together, it is not yet a heap. Ten, thirty, one hundred . . . How many before it crosses the threshold into a heap? Conversely, if you start with a heap and take a way one pebble at a time, when does it cease being a heap? Working with heaps means working with vague thresholds, paradoxical areas where things are and are not themselves. When does a place become a space? If I behold a place in a glance, and my hegemonic gaze surveys a space, where is the threshold between the glance and the gaze? If a glance takes half of a second, what would happen if I glanced for a whole second or two? At which point does my glance at a place turn into a gaze at a space? If a place is still a place if my experience is mediated through my perception, does it become a space if my perception is mediated with eyeglasses? What if it is mediated with binoculars, microscopes, and telescopes?

The point is simply this. Spaces are always already stranger than they appear. Less important than sorting out a typology of topology is the task of engaging in coexistential inhabitation in all of its paradoxes and ambiguities at all scales. In short, it is the task of becoming well rounded by becoming with others. Do not look from the outside and judge what counts as well rounded and what is pathologically misshapen, what counts as a living place or a lifeless space. There is no position outside, at least none that is not still inside. You are living in a global hothouse made up of an artificial continent of coziness surrounded by an ocean of poverty and devastation. Even if you go to Mars, your travel and outfits would be sponsored by the Crystal Palace.

Instead of attempting to take an outside position that judges spaces and places, be *in* the world. Instead of trying to escape, face the challenges of becoming well rounded, remaining exposed to the intimate proximity of things, including things you may not want to have in proximity, like capitalism, climate change, and mass extinction. That intimacy is depressing and horrifying. I do not want proximity with those things, but their surfaces are in my face. If coexistentialism is a form of optimism, it must be one that is optimistic about the worst (*pessimus*). If it is about a joyous affirmation of coexistence, it is a joy of dark ecology, a joy for which love and happiness are not opposed to mourning and grief. A glance at a place might be awash in crying and weeping, and in the blurry diffractions of tears there appears the radiant shine of things. That moment of intimacy with things is no solace for ecological emergency, but it does overcome the emergency of no emergency (Heidegger's *Not der Notlosigkeit*).[98] Coexistentialism does not make everything better, but the incessant drive to always make everything better—new and improved or safe and sound—is preventing anything different from emerging. It does not make the hurt, fear, and loathing go away. It lets the hurt hurt, it lets the emergency emerge, and it lets us be human.

NOTES

1. Jeff Malpas, *Heidegger and the Thinking of Place: Explorations in the Topology of Being* (Cambridge: MIT Press, 2012).
2. Heidegger, *Being and Time*, 140, 146.
3. Ibid., 79.
4. René Descartes, *Principles of Philosophy*, trans. Valentine Rodger Miller and Reese P. Miller (Dordrecht: Kluwer Academic Publishers, 1982), 42 (II 6).
5. Gaston Bachelard, *The Formation of the Scientific Mind*, trans. Mary McAllester Jones (Manchester: Clinamen Press, 2002), 86.
6. Gaston Bachelard, *The Poetics of Space*, trans. Maria Jolas (Boston: Beacon Press, 1994), xxxix.
7. Gaston Bachelard, *Water and Dreams: An Essay on the Imagination of Matter*, trans. Edith R. Farrell (Dallas: The Dallas Institute of Humanities and Culture, 1983), 5.
8. Bachelard, *The Poetics of Space*, 183–85.
9. Ibid., 203.
10. Karl Jaspers, *Von der Wahrheit* (Munich: R. Piper, 1947), 54. This translation is my own. The original German is as follows: "Jedes Dasein scheint in sich rund, jeweils geschlossen als dieses in seiner Welt, aber es ist un-geschlossen, weil nur durch anderes und in bezug auf anderes Dasein möglich, und weil selber in Verwandlung."
11. "There are children who will leave a game to go and be bored in a corner of the garret." Bachelard, *The Poetics of Space*, 16.
12. Ibid., xxxix.
13. Ibid., 234.
14. Ibid., 171.
15. Martin Heidegger, *An Introduction to Metaphysics*, trans. Ralph Manheim (New Haven: Yale University Press, 1987), 100.
16. Morton, *Realist Magic*, 21.
17. Ibid., 30.
18. Sloterdijk, *Bubbles*, 66.
19. Ibid., 340.
20. Ibid., 78.
21. Ibid., 27.
22. Ibid., 28. To what extent these "ecstatic beings" include nonhumans is an open question. He seems to follow Heidegger's restriction of the open to humans, but references to spheric dynamics in the climate, "self-organization" and "living" indicate a non-anthropocentric openness. Ibid., 24–28.
23. Jean-François Lyotard, *The Postmodern Condition: A Report on Knowledge*, trans. Geoff Bennington and Brian Massumi (Minneapolis: University of Minnesota Press, 1984), xxiv.
24. Slavoj Žižek, *On Belief* (London: Routledge, 2001), 15.
25. Foucault, *The Order of Things*. Edward Casey, *The Fate of Place: A Philosophical History* (Berkeley: University of California Press, 1997). Michael Hardt and Antonio Negri, *Empire* (Cambridge: Harvard University Press, 2001); *Multitude: War and Democracy in the Age of Empire* (New York: Penguin, 2004); *Commonwealth* (Cambridge: Harvard University Press, 2009).
26. Peter Sloterdijk, *Critique of Cynical Reason*, trans. Michael Eldred (Minneapolis: University of Minnesota Press, 1987).
27. Sloterdijk, *In the World Interior of Capital*, 3.
28. Ibid., 13. Cf. Sloterdijk, *Globes*, 46.
29. Sloterdijk, *Bubbles*, 62.
30. Ibid., 617.
31. Sloterdijk, *In the World Interior of Capital*, 9.
32. Ibid., 8.
33. Aristotle, *Metaphysics*, trans. Joe Sachs (Santa Fe: Green Lion Press, 1999), 1071b10.

34. Sloterdijk, *In the World Interior of Capital*, 8f.
35. Sloterdijk, *Globes*, 32–37.
36. Ibid., 79.
37. Sloterdijk, *Bubbles*, 66.
38. Ibid., 23.
39. Ibid., 25.
40. Nietzsche, *The Gay Science*, 181 (#125).
41. Sloterdijk, *Bubbles*, 24.
42. Sloterdijk, *Globes*, 144.
43. Sigmund Freud, *A General Introduction to Psychoanalysis*, trans. Joan Riviere (New York: Pocket Books, 1972), 296.
44. Sloterdijk, *Globes*, 71.
45. Sloterdijk, *In the World Interior of Capital*, 10.
46. Sloterdijk, *Globes*, 73.
47. Sloterdijk, *In the World Interior of Capital*, 25.
48. Ibid., 78.
49. Ibid., 13.
50. Sloterdijk, *Globes*, 25–26.
51. Sloterdijk, *In the World Interior of Capital*, 196.
52. Blake, *The Complete Prose and Poetry of William Blake*, 47.
53. Sloterdijk, *In the World Interior of Capital*, 169.
54. Fyodor Dostoevsky, *Notes from Underground*, 23–25.
55. Ibid., 17.
56. Sloterdijk, *In the World Interior of Capital*, 12.
57. Ibid., 196.
58. Ibid., 10–11.
59. Ibid., 161.
60. Ibid., 11.
61. Sloterdijk, *Bubbles*, 71. Foam (*Schaum*)—plural spherology—is the topic of the final volume in the *Spheres* trilogy. It has yet to be translated into English. Peter Sloterdijk, *Sphären III: Schäume, Plurale Sphärologie* (Berlin: Suhrkamp, 2004).
62. Johan Rockström et al., "Planetary Boundaries: Exploring the Safe Operating Space for Humanity," *Ecology and Society* 14.2 (2009).
63. Baird Callicott, *Beyond the Land Ethic: More Essays in Environmental Philosophy* (Albany: SUNY Press, 1999), 27, 43.
64. Douglas Christie, *The Blue Sapphire of the Mind: Notes for a Contemplative Ecology* (New York: Oxford University Press, 2013), 149.
65. Edward Casey, "Taking a Glance at the Environment: Preliminary Thoughts on a Promising Topic," in *Eco-Phenomenology: Back to the Earth Itself*, ed. Charles S. Brown and Ted Toadvine (Albany: SUNY Press, 2003), 188.
66. Edward Casey, *Spirit and Soul: Essays in Philosophical Psychology*, 2nd ed. (Putnam: Spring Publications, 2004), 327, 330.
67. Ibid., 328.
68. David Michael Levin, ed., *Modernity and the Hegemony of Vision* (Berkeley: University of California Press, 1993).
69. Casey, *Spirit and Soul*, 328.
70. Blake, *The Complete Prose and Poetry of William Blake*, 490.
71. Casey, "Taking a Glance at the Environment," 188f.
72. Casey, *Spirit and Soul*, 329.
73. Emmanuel Levinas, *Otherwise Than Being Or Beyond Essence*, trans. Alphonso Lingis (Pittsburgh: Duquesne University Press, 2002), 26.
74. Ibid., 114.
75. Ibid., 116.
76. Ibid., 88–93.
77. Casey, "Taking a Glance at the Environment," 190.

78. Emmanuel Levinas, *Entre Nous: Thinking-of-the-Other*, trans. Michael B. Smith and Barbara Harshav (New York: Columbia University Press, 1998), 80.
79. Levinas, *Otherwise than Being*, 88.
80. John Llewelyn is one of the most rigorous thinkers whose works contribute to carefully delimiting and extending the ethics of alterity articulated by Levinas and others, opening alterity to a wider distribution that includes all beings, human and nonhuman. John Llewelyn, *The Rigor of a Certain Inhumanity*.
81. Casey, "Taking a Glance at the Environment," 194.
82. Harry Heft, *Ecological Psychology in Context: James Gibson, Roger Barker, and the Legacy of William James' Radical Empiricism* (Mahwah: L. Erlbaum Associates, 2001).
83. James J. Gibson, *The Ecological Approach to Visual Perception* (Boston: Houghton Mifflin, 1979), 240.
84. Ibid., 307.
85. Ibid., 10.
86. Ibid., 307.
87. Ibid., 38.
88. Ibid., 23.
89. Gibson, *The Ecological Approach to Visual Perception*, 308.
90. Edward Casey, *Getting Back into Place: Toward a Renewed Understanding of the Place-World* (Bloomington: Indiana University Press, 1993), 314.
91. Snyder, *The Practice of the Wild* (Washington, DC: Shoemaker & Hoard, 1990), 41.
92. Bauman, *Religion and Ecology*, 128.
93. Haraway, *When Species Meet*, 7.
94. Morton, *Realist Magic*, 66.
95. Casey, *The Fate of Place*, 134.
96. Morton, *Dark Ecology*, 10.
97. Levinas, "Revelation in the Jewish Tradition," 192, 210n2.
98. On Heidegger's repeated phrase *Not der Notlosigkeit* (rendered as "distress of no distress"), see Heidegger, *Contributions to Philosophy (From Enowning)*, 75, 166, 168. See also Richard Polt, *The Emergency of Being*.

SIX

Interlude

A lot of what passes for environmentalism lacks musicality. Think of the term "sustainability." Why would anyone want to live in a society that aims to sustain its design and development? Imagine playing a piano with the sustain pedal pressed down. The notes just wash together. The envelope of a sound includes its attack and decay as well as its period of sustain and eventual release. Music without attack, decay, and release has no rhythm, and any melody or harmony cascades into an indiscernible, continuous fountain of noise—a kind of noise that would be unwanted in any genre. That kind of noise is not even found in noise music, which tends to use sounds that begin and end and not sounds that simply sustain. I am not at all saying that sustain has no place in music or that music that is noisy, saturated, or chaotic is bad. However, focusing exclusively or primarily on sustain is a failure to recognize the attack, decay, and release that mark the finite limits of sustain. The repeated refrains of "sustainability" and "sustainable development" are poorly composed songs that were written without concern for the finite conditions of actual coexistence. To the extent that they address the finitude of coexistence, they conceptualize finitude in terms of a compromise, which sounds something like this: We have been using energy and resources as if they were of infinite supply, and now we are learning to limit our energy and resource use, learning not to want so much. Living within limits in that sense means living out a compromised desire for limitlessness, a desire that resents the limits that thwarted the original plan for limitless growth. The death-denying desire for limitless resources is thus replaced with the death-denying desire for a limitless socioeconomic system that persists forever by staying within limits: permanent sustain, no decay and release.

Suppressing and resenting limits, finitude, decay, and death are ways of suppressing and resenting the conditions of ecological emergency. Suppression and resentment fail to let the emergency emerge. If there is a prime directive of the current period of unprecedented interconnectedness between humans and the land, life, air, and water of Earth, it is not to learn how to permanently sustain a limited civilization. Whoever we are, and that remains an open question, we do not need to learn how to sustain ourselves in the Anthropocene. We need to learn how to die in the Anthropocene.[1] We need to let the emergency emerge, which means becoming vulnerable to the overwhelming reality of other beings. It means remaining exposed to the climate catastrophe of globalization, which is like plunging into a blizzard.

In a blizzard, you experience the "*Ganzfeld effect*," which "renders here and there, up and down, foreground and background quite meaningless."[2] Your vision is rendered blind not by lack but by saturation. You are overwhelmed and cannot discern here from there. You are not surrounded, since there is no discernible "here" that could be surrounded by something "there." In the blizzard, what are you experiencing? It is not nothing, nor is it some vague "anything" that only turns into something specific when humans interpret it or society constructs it. It is the blizzard. It is exactly something, irremediably itself, overwhelmingly singular, getting in your face while remaining intangible. When it comes into contact with you, you find yourself overwhelmed by its disorienting strangeness, a nothingness which is not a lack or deficiency but a vertiginous radiance, an elemental givenness, an aesthetic magic in which the thing grasps you while remaining ungrasped. "So we are grasped," as Rilke says, "by what we cannot grasp."[3]

Grasped, enmeshed in the thick of its radiance, you cannot tell if what you are seeing is really the thing itself or an illusory appearance. You cannot be certain. There is no dotted line between here and there, appearance and reality, self and other, the mesh and the strange stranger. The challenge is not to make a dotted line but to learn to inhabit something to which philosophers have historically had a very bad allergy: the profound ambiguity and uncertainty of coexistence. Morton recommends "the manufacture of philosophical allergy medicines as a good alternative to the now defunct metaphysical dotted line business."[4] Coexistentialism is antihistaminic philosophy. It mitigates allergies to the ambiguity and uncertainty pervading the interconnected mesh of coexistence, including the ambiguity of what coexistentialism is and does. Out of the dotted line business, it perverts any clear distinction between what is inside or outside of coexistentialism, subject or object of coexistentialism, theory or practice of coexistentialism.

Coexistentialism, like a blizzard, overwhelms readily available distinctions.

When it makes distinctions, it often does so in a parodic parade, like the ridiculous multiplications of posts, in which coexistentialism becomes postsecular, posthumanist, postcolonial, poststructuralist, postnatural, postnihilistic, postfeminist, postdisciplinary, and posthistorical. With a little alliteration in a plurality of posts, coexistentialism makes up for the musicality lacking in refrains of sustainability. It is about attuning to the weird tunes of all beings, taking them to heart—recording, sampling, resampling, overdubbing, and mixing with the rest of them. It is about the birth of the Anthropocene out of the spirit of music. Along those lines—the staff, clefs, and ledger lines of musical notation—the current book can be understood as something like a series of musical notes from the coexistentialist underground. Among those musical notes, what follows is a brief interlude, a movement of play between (*inter-ludium*) the first half of the work, which introduced and exposed the intimate contours of coexistentialism, and the second half of the work, which elaborates on the themes of who we—"we"?—might become and what is to be done.

This is an intermedium, intermezzo. It is an interlude about interludes, which is to say, this interlude introduces a coexistentialist sense of liminality, the in-betweens, intermezzos, and intervals that make up the variegated textures and tunes of coexistence. This sense of liminality can be understood in different ways. It can be thought of as an updated version of the so-called golden mean espoused in ancient philosophies. For instance, in Confucianism, the Doctrine of the Mean (*zhongyong*) proposes that the best life is a life lived in the middle, in a path between extremes, although achieving such a mean is quite difficult and rare. "The Master said: 'Supreme indeed is the Mean as a virtue, but for a long time it has been rare among the people."[5] Buddhism likewise teaches that the middle way is best. This "middle" (*madhya*) is perhaps most emphasized in the philosophy of Nagarjuna, known for founding the Madhyamika ("middle way") school of Buddhism, where the middle is between asceticism and overindulgence, between nihilism and eternalism, between both/and and neither/nor.[6] Aristotle declares the middle or mean to be the site of excellence, avoiding the extremes of excess and deficiency. Ethically, this indicates that "virtue is a mean," involving habits, feelings, and activities that find the middle, such as courage, which is the virtuous mean between cowardice (deficiency) and recklessness (excess).[7] This carries over into politics. For Aristotle, "the political partnership that depends on the middling sort is best."[8] This entails that it is best for citizens to have a "middling" amount of property, avoiding the deficiency of poverty and the excesses of tyrannical appropriation. It also entails that "middling citizens" make the best legislators.[9]

The ancient sense of liminality needs to be updated for a contemporary ecological context, in which interdependence and uncertainty haunt every attempt to make a clear distinction between the middle and the extremes, the center and periphery. For instance, whereas Aristotle considered a middling amount of property to be virtuous, he also considered some humans to be property, which is to say, slaves. That means, at least in some cases, that "slavery is both advantageous and just."[10] The slavery, patriarchy, and provincialism of ancient societies do not perform the kind of interlude that is sought in coexistentialist contexts. Coexistentialist liminality is uncertain, ambiguous, and strange, without any determinate norms and prescriptions. It is a nonprescriptive performance of interconnectedness that opens possibilities for new modes of planetary coexistence. Along those lines, one can hear a coexistentialist interlude in Sloterdijk's theory of spheres, which he considers a generalized theory of media, a theory of intermediaries and mediations of messages and meanings, deliveries and receptions, translations and transferences, such that "media theory and sphere theory converge."[11]

A coexistentialist interlude can also be heard as a theory of what Gilles Deleuze and Félix Guattari call "interbeing," which is another name for their concept of the "rhizome" as a complex, nonhierarchical system. The rhizome cuts through dualisms with a "transversal movement" that proceeds "always in the middle, between things, interbeing, *intermezzo*."[12] "Interbeing" is also the name of a Buddhist monastic order that began in the 1960s under the leadership of Thich Nhat Hanh, a Vietnamese Zen Buddhist monk, for whom the sense of interconnectedness and interdependence expressed in the Buddhist doctrine of "dependent co-arising" (Pali, *paticcasamuppada*; Sanskrit, *pratityasamutpada*) can be described with the word "interbeing" (Vietnamese "*tiep hien*").[13] Interbeing is a messy mesh of strange strangers. However, the idea of interbeing can also suggest something that is not so messy and strange. It can turn into an average, neutral, lukewarm position that attempts to have its cake and eat it too, being in the middle of dualisms to enjoy the ambience without making an existential commitment. As Morton observes, there is a tendency among proponents of concepts like the rhizome to "wish to have it both ways," which risks "sheer nihilism" by absorbing the finely textured differences of coexistence into the neutral ambience of the between.[14]

In a coexistentialist milieu, being in the middle is not a lukewarm average. Movement picks up speed in the middle, calling for you to heighten your awareness in the blinding blizzard, to intensify your attention and quicken your reflexivity amid the profound ambiguity of coexistence, "ecoambiguity."[15] Consider the winged sandals of the Greek god Hermes (Mercury in Roman mythology), which are metonyms for his power to cross boundaries. As the god of transitions—messages and media—he does not stay in a lukewarm middle but picks up tremendous speed, like quicksilver (a synonym for liquid mercury). This speed is also represented in his winged hat and in the wings of his staff, the caduceus, up which two snakes wrap themselves, their faces meeting at the wings atop the staff. Consider three examples of inhabiting the ambiguities of the between in the midst of ecological emergency, where being in the middle is not a lukewarm average but an intensifying attention to intimate alterities. The first example regards knowledge, the second is about optimism, and the third is about so-called first-world problems.

First, knowing looks a lot like not knowing. For instance, some people think they know the best course of action to take to respond to ecological problems like air pollution, freshwater scarcity, environmental racism, and global warming. The prescriptions are diverse, including new laws and regulations, paradigm changes and worldview shifts, grassroots activism and community-building, technological development and design, science education and outdoor education, nature immersion and wilderness therapy, permaculture and community supported agriculture (CSA), taking shorter showers and becoming vegan, and so much more. Moreover, those are all prescriptions that consider ecological problems to be real. There are also prescriptions that claim the best course of action is to maintain business as usual and ignore the alarmism of environmentalists. For instance, consider critiques of "extreme environmentalism" or "ecofundamentalism," which posit extreme prohibitions against what is inaccurately described as religious extremist and fundamentalist agendas of environmentalism.[16] Such a critique achieves a trifecta of stupidity, as it misunderstands the theories and practices of ecology in relation to environmentalism, it overgeneralizes and cherry-picks phenomena of extremism and fundamentalism, and it does not account for the irony of its position—criticizing fundamentalism in support of another fundamentalism, market fundamentalism, which espouses strict and literal beliefs about the liberating power of the free market.

Whether prescriptions show concern for ecological issues or paranoid reactions against them, seeming to know what is best demonstrates the kind of imperial, unilateral perspective that ecology abhors. In a mesh of strange strangers, nothing is more delusional than seeming to know better than others, to know with any certainty what others should do. Seeming to know is delusion, and conversely, exposed to the uncertainty of one's relationship with other beings, unknowing resonates with gnostic attunement. The one best thing you can do for the environment is to start attuning to the weirdness of things and stop thinking that you can figure out a best thing you can do. What is the best thing to do to mitigate environmental destruction? Should I become vegan? Should I vote for politicians committed to environmental justice? Instead of trying to answer those questions, resonate with those things.

Second, nothing is more optimistic than critiques of Pollyannaish optimism. It is easy to criticize people for being unrealistically optimistic in the face of the volatile problems facing planetary coexistence. Imagine someone saying that the current mass extinction is not a serious problem, since there have been other mass extinctions, with five major extinction events having occurred before this one. Gaia will repair herself, and life will go on, just like it did before. Climate catastrophe? No worries, the resilience of Gaia will prevail. If it is not the resilience of Gaia, it might be the genius of the free market or technological innovation. Freshwater scarcity? No problem, new technological and market solutions will save the day. Kill the two birds of freshwater scarcity and sea-level rise with the one stone of water desalination plants. The idea that a resilient nature or an innovative culture can save the day fails to realize that the day, in many ways and for many beings, is already lost. Hundreds of millions of people already lack access to safe freshwater supplies. The abundance and diversity of species has already declined. The climate is already changed. Morton provides a good reminder, "everything is interconnected, and it sucks."[17]

Pollyannaish optimism warrants criticism and correction by more realistic perspectives. However, picking up speed in the middle of optimists and their critics, some ambiguities arise. One ambiguity is that purely Pollyannaish people do not actually exist. Even people who look at things entirely through the lenses of rose-colored glasses still see some shading in the world. Someone might write a book about the indomitable resilience of Gaia or the indefatigable growth of the free market, but their actual experience is doubtless not one that is simply and purely optimistic, with no sense of the immense suffering that comes with existing. For instance, in the theodicy of Julian of Norwich, she describes her visionary experience of the divine truth that "all things will be well," but just when you might think that she is articulating a naïve optimism, a closer look indicates that this phrase came to her through her relentless concern for the pervasive pain, suffering, sickness, and death that come with life as one of God's creatures, such that her optimism remains incomplete and dark, shrouded in divine mystery.[18] Even the most death-denying techno-optimist, who plans on never dying, knows that not everything is working out super well for everybody all the time.

My point here is not an *apologia* for optimism. If coexistentialism is an optimism, it is optimistic about the worst (*pessimus*), optimistic about sharing suffering and remaining vulnerable to the worst, exposed to the dust of this planet.[19] My point is that, the naïve optimists are not as naïve as their critics suppose, and the critics are more Pollyannaish than they admit. Yes, it is optimistic to think that everything will work out, but at least people who think that way are honest about being optimistic. Cynical criticisms of optimism do not own up to the optimism of their own cynical attitude, the optimism of their naïve certainty that one can simply rid oneself of Pollyannaish optimism if one chooses, as if the blind hope that things are getting better is like a product that you can avoid buying, as if I as an individual can listen to a good criticism and then stop thinking that everything could be well, stop thinking that the impossible is possible. What if that sense of the impossible is a constitutive aspect of caring for coexistence? I want so badly for all to be well. I hope it will be, and I think, perhaps, it could be. I do not think that because I am stupid, but because I care. In that sense, there is no non-optimism. There are just different kinds. Optimism that nature or God will fix everything, optimism that human ingenuity or rationality will fix everything, optimism that critique will fix the belief that everything can be fixed, and a coexistentialist optimism, which hopes not for the best but for what remains in the dust of this planet: just dust. It hopes against hope for the emergency to emerge.

Third, in various contexts, with particular profusion on social media, somebody will designate someone (oneself or another) as expressing "first-world problems," which generally refers to petty or trivial problems associated with life in a late capitalist society: when you have to settle for Pepsi instead of Coke at a restaurant, when the line to get a free piece of pizza in the office breakroom is long and moving slowly, or when your phone's battery is low and you have to go all the way into another room to get the battery charger.[20] The identification of first-world problems is itself another first-world problem. Complaining about people complaining about trivial problems is itself a trivial problem. Generally, designating something as a first-world problem is symptomatic of a cynical morality that looks at late capitalist society with a rolling eye, unimpressed by the decadence and corruption of its first-world milieu, which contrasts with the authenticity of the two-thirds world, where problems are noble and authentic, admitting of none of the pettiness and triviality that pervades late capitalist society. From this cynical perspective, those in so-called developed nations should be ashamed or at least self-deprecatingly amused by how excessively perturbed they are by the unimportant minutiae of their pampered lifestyles, by how much they lack engagements in real issues, and by how much they contrast with the authenticity of developing nations.

Cynics know very well that society has taken a wrong turn, and with that knowledge they escape their social milieu, observing their sphere from the outside, reaching transcendent heights from which to look down on society in judgment or, with rolling eyes, to simply look up and away into the nihilism of whatever. Nihilism is an easy escape from the finely textured triviality and pettiness that are woven throughout the mesh of coexistence. Annoyance that a thing you want is a little bit further away from you than you first thought it was is not restricted to people in developed nations. Whether that thing is a phone charger or a candle does not make a world of difference. The attempt to make a dotted line between them is just a cynical way to avoid the mess of the mesh. Everybody deals with small details in their lives that become tedious or annoying, because living with strangers is not always fun and games, nor is it always monumental or grandiose. The problem is not that people like me in the first world focus too much on the small and trivial and not enough on larger, more important issues. The problem is the dotted line between the small and the larger, the trivial and the important. On the smallest scales, practices of becoming-with are practices of becoming worldly. The beating heart of intimacy has microspheric and macrospheric pulses. Rather than casting a cynical eye toward your phone charger, you could consider how to sincerely engage in the practice of attending to the things you exist with. I am not saying do not complain when autocorrect misspells something in your text message. That would just be piling cynicism upon cynicism. When you complain, let it hit you like a blizzard—living with autocorrect and text messages, you are in profoundly close contact with strangers, for better, for worse, and for boring.

The minutiae of everyday life are big issues, and any responses to large-scale problems like climate change are still going to involve petty disappointments, like not having comfortable leg room on a flight to a climate change conference. If the world's political leaders were signing a document guaranteeing peace, justice, and flourishing for all, it would still be annoying if the pen ran out of ink and had to be replaced with another pen that was not very stylish. Everything is interconnected, and it sucks, even the trivialities, perhaps especially the trivialities. Criticisms of first-world problems are a red herring from the sort of problems that are really unique to the first world, problems involving massive overconsumption and unjust distributions of resources. The uniquely first-world problem is not about getting annoyed with the trivialities of affluence. It is the production of that affluence in the first place. Getting more concerned, annoyed, and worried about the minutiae of everyday life could, perhaps, facilitate more just and sustainable ways of sharing in planetary coexistence.

Could any of this interlude make sustainability sound a little more musical? If so, it might sound something like Jeffrey Bell's account of the "makings of a sustainable life." For Bell, "Aristotle's middle constitution may be seen to represent a similar effort" to Deleuze and Guattari and to Henri Bergson, who likewise emphasize movement through the middle.[21] For Bergson, the movement of the middle is not a lukewarm average but a tremendous and transformative speed.

> The mind has to do violence to itself, has to reverse the direction of the operation by which it habitually thinks, has perpetually to revise, or rather recast, all its categories. But in this way it will attain to fluid concepts, capable of following reality in all its sinuosities and of adopting the very movement of the inward life of things.[22]

What Bergson, Deleuze, and Guattari have in common with Aristotle is a commitment to a middle constitution in which the state and the people—virtuous people, media people—avoid falling into two contrary forms of violence or excess. The first is the violence of a desire for unlimited acquisition and consumption ("nomadic violence"), and the second is the violence of the moral stifling and political suppression that seek to limit the desire for unlimited acquisition ("institutional violence"): "A state that is stable and yet capable of adapting to changing conditions will thus need to move between the two modes of violence, or it will be a state at the edge of chaos."[23] The challenge then is to move in between.

> If one can strike a dynamic state that swerves neither too much toward the excesses of unlimited desire, nor too much to the stifling restrictions of life-denying simplicity, then one achieves a good life as an individual, or, in the case of the middle constitution, one achieves a good, self-sufficient state.[24]

This sustainable life avoids unjust overconsumption without resorting to the stifling and suppressing desire for pure sustain that resonates in so many refrains of sustainable development, sustainable design, and sustainability transitions. Less about sustain and more about intermezzos, this sustainability plays in the middle of things, carrying a tune and being carried away by the tunes of others, attuning to intimate alterities enmeshed in ecological emergency.

NOTES

1. Roy Scranton, *Learning to Die in the Anthropocene: Reflections on the End of a Civilization* (San Francisco: City Lights Books, 2015).
2. Timothy Morton, "Elementality," in *Elemental Ecocriticism: Thinking with Earth, Air, Water, and Fire*, ed. Jeffrey Jerome Cohen and Lowell Duckert (Minneapolis: University of Minnesota Press, 2015), 272.
3. Rilke, "A Walk," 176.
4. Morton, "Elementality," 279.
5. Confucius, *The Analects*, trans. Raymond Dawson (Oxford: Oxford University Press, 2008), 6.29.
6. David J. Kalupahana, *Nāgārjuna: The Philosophy of the Middle Way* (Albany: SUNY Press, 1986). For a translation of the major work of Nagarjuna, *Mūlamadhyamakakārikā*, see Nagarjuna, *The Fundamental Wisdom of the Middle Way*, trans. and comm. Jay Garfield (Oxford: Oxford University Press, 1995).
7. Aristotle, *Nicomachean Ethics*, trans. Joe Sachs (Newburyport: Focus Publishing, 2002), 1107a.
8. Aristotle, *The Politics*, trans. Carnes Lord (Chicago: University of Chicago Press, 1984), 1295b.
9. Ibid., 1296a.
10. Ibid., 1255a.
11. Sloterdijk, *Bubbles*, 31.
12. Gilles Deleuze and Félix Guattari, *A Thousand Plateaus: Capitalism and Schizophrenia*, trans. Brian Massumi (Minneapolis: University of Minnesota Press, 1987), 25.
13. Thich Nhat Hanh, *Interbeing: Fourteen Guidelines for Engaged Buddhism* (Berkeley: Parallax Press, 1998), 27.
14. Morton, *Ecology without Nature*, 54.
15. Scott Slovic, Swarnalatha Rangarajan, and Vidya Sarveswaran, eds., *Ecoambiguity, Community, and Development: Toward a Politicized Ecocriticism* (Lanham: Lexington Books, 2014).
16. Rögnvaldur Hannesson, *Ecofundamentalism: A Critique of Extreme Environmentalism* (Lanham: Lexington Books, 2014). See also Patrick Allitt, *A Climate of Crisis: America in the Age of Environmentalism* (New York: Penguin Press, 2014). Like Hannesson,

Allitt presents an optimistic portrayal of a growing free market and an alarming rebuke against what he perceives as a culture of alarmism that is exemplified in the antinuclear movement, the rise of the counterculture in the 1960s, critiques of overpopulation, and the environmental movement.

17. Morton, *The Ecological Thought*, 33.

18. Julian of Norwich, *Showings*, qtd. in Rosemary Radford Ruether, *Women and Redemption: A Theological History*, 2nd ed. (Minneapolis: Fortress Press, 2012), 87.

19. I allude to Eugene Thacker (*In the Dust of this Planet*), as his horror of philosophy trilogy articulates this kind of affirmation of pessimism. "If pessimists are enthusiastic about anything, it is about *pessimus*, 'the worst,' resplendent in every possible outcome, glimmering around every corner, an *ecstasy of the worst*." Thacker, *Starry Speculative Corpse*, 156f.

20. Streeter Seidell, *White Whine: A Study of First-World Problems* (Avon: Adams Media, 2013).

21. Jeffrey A. Bell, "Undoing the Subject: Deleuze and the Makings of a Sustainable Life," in *New Directions in Sustainable Design*, ed. Adrian Parr and Michael Zaretsky (New York: Routledge, 2011), 234.

22. Henri Bergson, *An Introduction to Metaphysics*, trans. T. E. Hulme (New York: G. P. Putnam's Sons, 1912), 69f.

23. Bell, "Undoing the Subject," 234. In Aristotle's *Politics* (1257a), he distinguishes between a natural kind of property acquisition that is limited to needs and household affairs and an unnatural acquisition of property, the commercial art of acquisition (*chrematistike*; money making).

24. Bell, "Undoing the Subject," 238.

SEVEN

After Humanism

An outcry against the excessive growth of the human population has been a common refrain in environmentalism for decades, exemplified in works like Paul Ehrlich's *Population Bomb*.[1] Overpopulation is sometimes singled out as the leading driver of the global environmental crisis. There are simply too many of us humans around. Ideas of who exactly is responsible for this "too many" reveal racist, classist, and sexist assumptions, revolving around a typically unspoken assumption that too many women of color are having too many babies in developing nations. An important corrective to the concern with overpopulation diagnoses contemporary environmental problems not primarily in terms of overpopulation but overconsumption and unjust distribution. For instance, in presenting his "social ecology," Murray Bookchin puts it this way. Even if the human population was extremely small, even around ten million people total, those people would nevertheless produce a global environmental crisis if they were committed to capitalism. This is the function of "capitalism with its competitive 'grow or die' market economy—an economy that would devour the biosphere whether there were 10 billion people on the planet or 10 million."[2] That might seem hyperbolic, but if that very small population of capitalists learned that they could profit off of disasters (i.e., "disaster capitalism"), then what might have seemed hyperbolic begins to appear inevitable, as it is a matter of time.[3] The real threat to planetary coexistence is not the sheer number of human beings. It is the humans living in the Crystal Palace of capitalism that are endangering and extinguishing life on Earth.

To argue that overconsumption and unjust distribution are causes of ecological devastation is not to say that overpopulation has nothing to do with it. Conceivably, in Bookchin's example of a small human population, it would take those humans somewhat longer to consume and destroy the same amount as a population of a few hundred million or a few billion capitalists. In any case, the ways that resources are distributed and consumed are crucial factors contributing to current ecological problems. Overpopulation is not the sole cause of the problems in the world today. Indeed, in some sense, the critique of overpopulation has not just been too narrow, focusing on population while ignoring consumption and distribution. Rather, the critique of overpopulation has been completely backward. Consider the possibility that the excessive growth in the number of human bodies in the last century is not accurately described as overpopulation but as underpopulation. Make no mistake: I am not advocating a pro-natalist ideology. If I am advocating anything, I am advocating coexistence.

You could think with Sloterdijk and see that where there would be a well-rounded people—a *demos, populus,* or *Volk*—there is only foam, a population in the quantitative sense but not in the spherological sense. You could think with Haraway and see that there are a lot of human babies on Earth, but not enough companion species, not enough kin. "So, make kin, not babies!"[4] If people are not just isolated individuals but are constituted in intimate relationships, then there are not enough people on the planet right now, despite the astounding proliferation of humans throughout the last one hundred years, growing from less than two billion to more than seven billion humans. Furthermore, if people are not just human persons but are all persons in an animistic sense that includes all things, then there are not enough people on the planet in the sense that countless beings are having their personhood systemically denied, excluded, and destroyed, including many human persons who do not fit the wealthy, heterosexual, white, able-bodied, neurotypical, male standard of humanity. Considering this problem of underpopulation, it would be a mistake to frame population issues in terms of a need to decrease the number of people on the planet or even the number of human people in particular. The population problem is not a numbers game. It is about intimate alterities. The challenge is to let people be people and not efface or rob them of their personhood. This means letting humans become human.

Coexistentialism is after humanism in the same way that it is after God and nature, subsequent to humanism and its ideas of human nature while also seeking an affirmation of what it means to be human. After humanism, there remains more to follow. Remaining exposed to strange strangers during an epoch of ecological emergency, humans are called to rethink their existence, which means rethinking their coexistence in a world full of persons. Existential approaches to phenomenology and to the study of religion provide renewable resources for thinking through the intimate intertwining of humans with a world full of persons, thinking through what can be called anthropocosmology. This anthropocosmology is complicated by the strange beings populating the Anthropocene, things Morton calls "hyperobjects" due to their massive distribution in space and time, things like global warming, Styrofoam, capitalism, plastic bags, Facebook, and the Big Bang. Under the weight of hyperobjects, the human of anthropocosmology is rendered weak, lame, and hypocritical, which is only bad news if you are attached to the excessively armored ego of classical humanism. Following the outlines of an anthropocosmology and its openings for weak, lame, and hypocritical humans, I consider how this sense of human existence provides a radical role for the humanities in the form of ecological humanities.

ANTHROPOCOSMOLOGY

Who am I? That is one of the compelling questions of existentialism, particularly because existentialists are seeking to live that question, to make it personal. Existential truth is "a truth which is truth *for me*."[5] How can you really live a question? Who am I? What am I doing here? I know you know who you are, and I know that you can ask and answer questions about who you are: questions regarding where you live, questions about whether you believe in God, questions about definitive moments or relationships in your life, questions about your goals or your special purpose, and so on. Asking questions is one thing, to bear them existentially is quite another. What am I doing here? The question comes with a shock, horrifying but not necessarily unpleasant. You lose your grip. "I" is in suspense, and things close by shimmer with nothingness—not nothing, but the nothingness whereby a thing is so irremediably itself that you cannot make anything of it, the emptiness whereby things grasp you while you remain empty handed. There are no answers here, no meanings, no nothing, only the open question. Who am I? This is not the same as the audacious attempts at self-knowledge in Enlightenment humanism or secular humanism or any humanism that pretends to know what it means to be human. If existentialism can be defined as a kind of humanism, as Sartre claims, it is a radically open humanism. It must be an abysmal humanism, a dark humanism, a humanism without humans in the sense that there is no presupposed essence that defines human existence ("existence precedes essence").[6] However, in that openness, perhaps humanism too must remain in suspense.

Part of the problem with humanism is the "ism." As Heidegger observes, "'-isms' have for a long time now been suspect. But the market of public opinion continually demands new ones. We are always prepared to supply the demand."[7] The "ism" is a general problem that all thinking faces as it attempts to remain engaged and not fall back on ossified categories. It is not the primary problem with humanism. The human is also a problem. If thinking is thinking of being, in the double sense of the genitive—thinking about being and from being—then thinking is not oriented around the human. It is oriented around the clearing in and as which the human being comes to be, the dwelling or abode of human existence, the spheres of being-in. In that sense, existentialism is not a humanism, and indeed, it is not even an existentialism if what one means by existentialism involves a presupposition about the relationship between essence and existence. Accordingly, Heidegger disagrees with Sartre's formula that existence precedes essence. For the ecstatic opening of being-in-the-world, the difference between existence and essence remains an open question.[8] That openness is foreclosed in the materialism that treats humans like biological objects, the idealism that treats humans like subjects correlated to a world of objects, and the pragmatism that treats humans like practical decision-makers. Heidegger is not giving up the quest for humanity. He is situating it in the open, so that the question of humanity is a question not of a transcendental subject but of an exposure to the world.

Merleau-Ponty similarly parts with Sartre's humanism in favor of a more radical openness to the world. In his early work, Merleau-Ponty's "radical reflection" is not very different from Sartre's existential phenomenology. It holds itself in suspense to disclose the un-reflective, pre-thematic life of consciousness, the life of the existential subject is that of "a subject destined to the world," "a process of transcendence toward the world."[9] Assigning a primary role to perception and the sensing body, Merleau-Ponty understands being-in-the-world as "body-in-the-world."[10] This body-world connection deepens throughout Merleau-Ponty's career, leaving behind the subject-object correlation to entangle humans in the stranger mesh of a more elemental flesh. In his unfinished manuscript *The Visible and the Invisible,* he speaks of a style of "hyper-reflection" (*sur-réflexion*), which does not lose sight of or efface the things and perceptions of brute, wild, pre-thematic being, attending to "the secret of our perceptual bond" with the world, "our mute contact with the things, when they are not yet things said."[11] Our bond is the flesh (*la chair*)—"a *general thing,* midway between the spatio-temporal individual and the idea, a sort of incarnate principle that brings a style of being wherever there is a fragment of being. The flesh is in this sense an 'element' of Being."[12]

The flesh brings the style of the intertwining (*l'entrelacs*) to the world, the style of the chiasm (*le chiasme*), wherein the sensing and the sensed crisscross and intertwine, a reversibility that brings self and other, perception and meaning, visible and invisible, matter and form, to appear together with one another—overlapping, transgressing, transcending, and relating to one another.[13] The human body is intertwined with the world, "made of the same flesh as the world."[14] However, this intertwining should not be mistaken for a connective continuity without difference, as some phenomenologists seeking nature immersion tend to think.[15] Any continuity of coexistence is an emergent effect of the differences between things. Accordingly, reflecting on structuralist understandings of language as a system of diacritical differences in which the meaning of any sign depends on its relation with other signs, Merleau-Ponty describes a "unity of coexistence" by likening it to the unity of "the sections of an arch which shoulder one another."[16] The unity of coexistence is differential. This implies that there is never a complete coincidence of the body and the world, touching and the touched, sensing and sensed. Rather, "the coincidence eclipses at the moment of realization."[17] The reversibility of the flesh is never fully realized in any fact. The flesh realizes itself partially in the "coiling over of the visible upon the seeing body," opening up "by dehiscence or fission of its own mass" a space within which the body can sense itself and the flesh of the visible world.[18] Being entangled in flesh means being in the open—exposed to the unfathomable closeness of beings.

Like Heidegger, Merleau-Ponty avoids the subjectivism of Sartre's existential phenomenology by resituating human existence in the open. Unlike Heidegger, for whom only humans have a share in the opening of the world, Merleau-Ponty's flesh has openings for nonhumans. The world-forming agency described as intentionality, consciousness, or *logos* is not exclusively human. There is a *logos* folded into the flesh of the natural world.[19] Heidegger's human is worldlier—intimately in and with—than Sartre's, and Merleau-Ponty's nonhumans are worldlier than Heidegger's. Merleau-Ponty's works have figured into the efforts of many eco-phenomenologists and environmental philosophers to challenge anthropocentrism and open philosophical inquiry up to more carnal modes of participation in the cosmos, attending to the *logos* crisscrossing the flesh of my body and the flesh of the world.[20] Attending to that *logos*, Merleau-Ponty's philosophy is anthropocosmology. "Philosophy is nature in us, the others in us, and we in them."[21] This is not simply to say that he is non-anthropocentric. Rather, the intimate intertwining of the flesh complicates any mutually exclusive opposition between centers and peripheries. Affirming a differential unity of coexistence and redistributing *logos* across the flesh of the world, Merleau-Ponty's anthropocomology expresses intimations of an ecological existentialism or coexistentialism.[22]

Coexistentialism is even more explicitly articulated in the works of Bachelard, from whom the term "anthropo-cosmology" comes.[23] While Merleau-Ponty accounts for the flesh as an element of being, Bachelard's writings plunge much more deeply into the specificities of elemental being, accounting for the singular plurality of elements (earth, air, water, fire) and the pull that their uncanny agencies have on the poetic imagination. Even something as seemingly simple as a candle's flame compels Bachelard into an extended meditation. Even a "flame is a world."[24] As a world, provoking the poetic imagination, the flame harbors its own alluring agency or value. As "the world gravitates about a *value*," the flame is a world and, indeed, the flame is the center of a world, not merely a worldly periphery or horizon waiting for a human center to participate in its potential meaning. In anthropo-cosmology, every being is a center of a world. This is not only true of humans and elemental forces like flames and rivers. Even everyday household objects are the centers of worlds. "The evening lamp on the family table is also the center of a world."[25] The anthropocentrism in existentialism thus gives way to a vertiginous profusion of centers in coexistentialism. "In this coexistentialism every object invested with intimate space becomes the center of all space. For each object, distance is the present, the horizon exists as much as the center."[26]

Anthropocosmology is being-in-the-world unchained from human exceptionalism. For Bachelard, every being is the shepherd of being, the custodian of the cosmos. Articulating being-in-the-world by using any familiar thing as an entry point or symbol, a phenomenological description of a bottle can show how "a paltry thing becomes the janitor of the world."[27] Furthermore, unchained from human exceptionalism, anthropocosmology also unchains being-in-the-world from the lack of imagination typical of many philosophers. Bachelard mentions that formulas like "being-in-the-world" are "too majestic" for him, such that they are not real matter of experience but are more indicative of the tendency of metaphysicians to group everything that is not themselves into a big surround called "World." The term "anthropocosmology" is misleading. Bachelard is less interested in formulas about worldhood than he is in nests, shells, miniatures, attics, basements, meadows, forests, mountains, rivers, and a solitary flame. A vast space in the cosmos is not experienced with images of world, but images of "inner immensity" or "intimate immensity."[28] The point of anthropocosmology is not to reduce things to a human-world correlate. The point is to situate humans within the fabric of a multiplicity of things, each of which is its own unfathomable world, making a world of difference. To "neglect nothing of the anthropo-cosmic tissue of a human life" means to attend to the multifarious nonhuman or inhuman worlds intersecting with human worlds.[29] It requires breaking humans out of the lonely castles of subjectivism and humanism, exposing them to their cohabitation with intimate alterities. In short, it requires making humans well rounded (see chapter 5, "Roundness").

By what appears to be independent invention, Bachelard's anthropocosmology is echoed by the historian of religions Mircea Eliade.[30] For Eliade, the academic field of religious studies has an existential force, disseminating images and symbols that compel modern humans to change their lives, to become anthropocosmic. Eliade indicates how modern humans can benefit from understanding anthropocosmic symbolism through inquiry into the history of religions.

> Still with the aid of the history of religions, man might recover the symbolism of his body, which is an anthropocosmos. What the various

> techniques of the imagination, and especially poetic techniques, have realised in this direction is almost nothing beside what the history of religions might promise. . . . By regaining awareness of his own anthropocosmic symbolism—which is only one variety of the archaic symbolism—modern man will obtain a new existential dimension.[31]

Eliade thus expands the horizon of potential techniques for cultivating attention to the anthropocosmic tissue of coexistence. The poetic techniques cherished by Bachelard are the tip of the anthropocosmic iceberg. For thousands of years, the religious traditions of the world have been developing techniques for rejuvenating the ecstatic openness whereby humans are intertwined with intimate alterities.

Numerous rituals, symbols, scriptures, and beliefs are oriented toward rendering human worlds porous to other worlds. This is evident in "a whole system of micro-macrocosmic correspondences" that shows up throughout the history of religions, as can be seen in images of "the assimilation of the belly or womb to a cave, of the intestines to a labyrinth, of breathing to weaving, of the veins and arteries to the sun and moon, of the backbone to the *axis mundi*."[32] These cross-cosmic correspondences introduce the sacred into the "various existential situations" of human beings, thus facilitating "the sacramentalization of physiological life."[33] Introducing the sacred into existential situations means leading humans into encounters with that which is "wholly other" (*ganz andere*), a "numinous" alterity.[34] Eliade is following Rudolf Otto's definition of the sacred or the holy (*das Heilige*) as a numinous or wholly other power characterized as *mysterium tremendum et fascinans*—an inexplicable mystery encountered in horrifying tremors and shuddering dread (*tremendum*) as well as wondrously rapt fascination (*fascinans*).[35]

The sacred is dreadfully fascinating and wondrously uncanny. It can show up in a variety of anthropocosmic correspondences: when an Aboriginal Australian tells a story about the Dreamings that produced the kinship between humans and dingoes; when an astrologer interprets the alignments of the planets at the time of your birth (a natal chart); when a Christian touches holy water, which corresponds with the baptismal waters of Jesus, the waters from the beginning of God's Creation (Genesis), and the waters of life promised to the faithful after the end of the world (Revelation); when a practitioner of Yoga aligns her spine with the center of the world, Mt. Meru; when a taiji practitioner harmonizes her flow of energy with the energy of Earth and the cosmos. In anthropocosmic correspondences, the human is grasped by the ungraspable, finding itself by losing itself in the emptiness of things, empty handed. This anthropocosmic style is expressed by Dōgen, the thirteenth-century Japanese Buddhist who founded the Soto school of Zen. "To study the buddha way is to study the self. To study the self is to forget the self. To forget the self is to be actualized by myriad things. When actualized by myriad things, your body and mind as well as the bodies and minds of others drop away."[36] In brief, "what buddhas call the self is the entire universe."[37] If anthropcosmology had a slogan, that could be it. Anthropocosmology could also find a slogan in a regularly recited refrain of the Western esoteric tradition, a refrain that appears in *The Emerald Tablet* (*Tabula Smaragdina*), a text attributed to the legendary Hermes Trismegistus. The text formulates a maxim of anthropocosmic correspondence: "That which is above is like that which is below, and that which is below is like that which is above."[38] As above, so below; as goes the macrosphere, so goes the microsphere.

While Bachelard and Eliade did not explicitly address environmental issues, the anthropocosmic visions of religious traditions can be understood as an alternative to the false dichotomy between anthropocentric and non-anthropocentric environmental ethics, both of which presuppose a mutually exclusive opposition between center and periphery, an opposition rendered untenable by intimate coexistence in the mesh of strange strangers. One of the most thorough applications of an anthropocosmic vision to a religious environmental ethic comes from Tu Weiming, who finds an anthropocosmic vision—a vision of the "unity of Heaven and humanity" (*tianrenheyi*)—in Confucianism, which emphasizes the cultivation of the self in intimately interwoven personal, social, ecological, and cosmic spheres of coexistence.[39] Tu is not saying that everybody should learn from Confucianism, but he is saying that the modern notions of a rational human subject need to be corrected for the sake of humanity and the future of life on Earth. If you do not find religious techniques for cultivating an open, anthropocosmic sense of humanity useful, perhaps phenomenological attention or poetic imagination will work. There is no shortage of techniques for stepping outside of oneself, finding your humanity by putting your humanity in question, becoming human by opening up to intimate alterities.

TANGLING WITH HYPEROBJECTS

If coexistentialism is anthropocosmic, it is neither subjectivistic nor anthropocentric. Does this mean that it is a posthumanism? It is a form of posthumanism in the sense that it pulls humans into the orbits of nonhuman agencies, rendering humans abominably eccentric. It is a form of posthumanism in the sense that humanism is rendered untenable, as ecological emergency has burst the bubble of the lonely individual human—"a skin-encapsulated ego."[40] Coexistentialism can be considered a form of posthumanism, and indeed, it can be thought of with a multiplication of posts: postsecular, postcolonial, poststructuralist, postnatural, postnihilistic, postfeminist, and postdisciplinary. However, going postal sounds reactionary. When it feels impossible to bear the ambiguities of the Anthropocene, it can seem convenient or even necessary to draw a clear and distinct boundary—a fence made of posts—separating those who get it (posthumanists) from those who do not (modern humanists or premodern traditionalists). Ironically, the desire for transcendence and newness in postal fashions are actually very old. Postal fashions are severely behind the times, as Slavoj Žižek quips.

> When G. K. Chesterton describes his conversion to Christianity, he claims that he "tried to be some ten minutes in advance of the truth. And I found that I was eighteen years behind it." Does the same not hold ever more for those who, today, desperately try to catch up with the New by way of following the latest "post-" fashion, and are thus condemned to remain forever eighteen years behind the truly New?[41]

There is a triumphalist either/or morality resonating in the assertion that a posthuman condition has been achieved or that anyone is living a posthuman life. Posthuman sounds like it is human and then some, updated, maybe bigger and better, perhaps new and improved, smarter than human, smarter than humanism, with new technological and ecological accessories and add-ons. If we are no longer in the grips of the philosophical suppositions of humanism, it is not because we have gotten beyond those suppositions but because their failure has put us back in the beginning, square one—not knowing who we are. That is where anthropocosmology takes place, not in the position of one who knows better, but in a disposition that is open to intimacy with the radical alterity of things, human, nonhuman, or otherwise. In that sense, anthropocosmology is a form of what John Caputo calls "radical hermeneutics," which does not seek a postal position but instead celebrates the disposition of "not knowing who we are."[42] The uncertainty is a constitutive dimension of being a person. "Being a person," as Morton points out, "means never being sure that you're one."[43] Coexistentialism celebrates the uncertainty of being human amid a profusion of intimate alterities, in which every bit of every other radiates the horribly fascinating allure of the wholly other. As Derrida says, *tout autre est tout autre* (every other is wholly other).[44]

To be human means always trying to find our humanity, to become anthropocosmic. The wholly other always remains to come, unheard of (*absurdus*). If it was present, it would not be wholly other. If it could be spoken of, it would not be wholly other. "For the *tout autre* would not be *tout autre* were the arrows of language to reach that far; the *tout autre* must always be to come, unheard of and unthought."[45] Posthumanists have heard of and thought of a lot, including Derrida's *tout autre*. Coexistentialists hear and think with posthumanists, but they also attune to the detuned tune of that which remains unthought and unheard of. If they take up phenomenology, it is "a phenomenology without phenomena."[46] They tell ongoing stories about learning to play with strangers, stories of companion species. Coexistentialism, like Haraway's feeling for companion species, maintains an "effort to be in alliance and in tension with posthumanist projects."[47] In other words, companion species are "inoculated against post-humanism but rich in com-post," where *com-post* marks an alliance *with-post*humanism, an alliance in becoming "inoculated against human exceptionalism but rich in humus, ripe for multispecies storytelling."[48] Similarly, Morton's ecological thought is inoculated against human exceptionalism while also "seeping through posthuman ideological barriers" that dismiss the human or absorb the human into some vague holism of processes and networks.[49]

Coexistentialism is about composting the human in anthropocosmic ties with intimate alterities. It is less of a humanistic or posthumanistic thought and more of an ecological thought, which ends up providing a more accurate sense of what it means to be human. It is about finding oneself while opening oneself to the uncanniness of ecological interconnectedness, a mesh of strange strangers. "What if being human is the encounter with the strange stranger—in other words, at a certain limit, an encounter with the inhuman? Isn't this the very 'posthumanism' for which some are yearning?"[50] Every strange stranger is completely strange; every other is wholly other. Every single thing in the mesh harbors an irreducible alterity that never appears, or more accurately, every "other appears as a being whose appearance appears without appearing."[51] Every other appears, shows up, makes contact, yet what appears is so irreducibly itself that all of its appearances appear without appearing, which is to say, they appear as a spectral shimmer of what remains to come. It touches you while remaining intact. Completely in your face, the other appears without being present. As Derrida says, the other is a "newcomer" or "guest" (*arrivant*), always remaining "to come" (*à-venir*), in "the future" (*l'avenir*), appearing by haunting and distorting the present with its compelling insistence on justice—arriving as itself, just itself, wholly other.[52] The phrase "strange stranger" is Morton's way of translating Derrida's *arrivant*.[53]

While Levinas and some other proponents of alterity-oriented ethics restrict alterity to humans, anthropocosmology plunges you into a world in which every other is wholly other. Coexistence means being intimately intertwined with countless strange strangers, not just humans or seemingly natural things like plants, animals, bacteria, and fungi. Even the lamp on a table, the flame on a candle, and the light inside of the refrigerator each insist on justice. As Alphonso Lingis recognizes, this extended sense of Levinasian alterity suggests that each thing "clamors for all our attention," and indeed, the whole layout of the landscape "weighs on us as an imperative."[54] The alterity of the face is harbored in every surface, with imperatives issuing from the insistent arrival of all things.

> The key, the inner formula of a mango, a willow tree, or a flat smooth stone, is never grasped; the real thing is before our perception as a task

> for an exploration. But the real thing is not the sum of all that we have recorded of it; it closes in upon itself, remains exterior, always beyond all that our perceptional samplings have turned up of it, not a given but an external ordinance. The reality of things is not given in our perception, but orders it as an imperative.[55]

Opening up to the strangeness of human strangers is immensely challenging, and anthropocosmology intensifies and multiplies that challenge by pointing out that strangeness far exceeds human beings and extends through everything. Every bit of other is every bit other. Every single thing in the mesh of coexistence is a strange stranger.

To make things more complicated, like never before in history humans have become intimately interwoven with Earth's systems, so much so that the geological epoch that has characterized the last twelve thousand years is now ending. The Holocene has given way to a new geological epoch or boundary event, the Anthropocene, in which human impacts are evident throughout Earth's systems of life, land, air, and water.[56] Although this new geological epoch is named after the human (*anthropos*), it is not because the human is in charge; quite the opposite, the idea of the human as the apex or sole proprietor of meaning or as the conqueror of nature or master of the universe has become untenable. Humans are shaping the chemistry of the atmosphere and the oceans, diverting rivers, altering DNA sequences, patenting organisms, spreading toxics, plastics, metals, and radioactive material around the planet, yet we are not the special agent in charge. "*'Anthropocene' is the first fully antianthropocentric concept.*"[57] We are under the weight of massive unintended and unexpected consequences of human actions. Not all humans are or were equally involved, as participation is differentiated along multiple lines of different subject positions. Nonetheless, these are consequences of human actions. Climate change, ocean acidification, freshwater scarcity, antibiotic resistant bacteria, biodiversity loss, and pollution are things for which humans are responsible. They are clamoring for our attention, weighing on us as an imperative.

A defining characteristic of this period is the depositing of a layer of carbon around the planet due to human extraction and consumption of fossil fuels. Not to be outdone, radioactive material found its way into this layer through nuclear weapons testing and nuclear power generation. Plastic and Styrofoam are part of this layer as well. These layers seem to originate somewhere between the industrial age, perhaps with the invention of the steam engine, and the nuclear age, with the invention of the atom bomb. The exact dating of the beginning of the Anthropocene is not my concern here. My concern is how the Anthropocene is complicating the challenges of becoming anthropocosmic. Of course, humans have always been altering the conditions around them, but clay pottery, cave paintings, and agriculture have a different spatiotemporal distribution than the weird things populating the Anthropocene, things like plastic, Styrofoam, the Internet, nuclear power plants, and global warming. Those things are massive. Plastic is not just in a few places here and there. It is part of a stratigraphic layer of Earth's crust. The burning of fossil fuels is changing the chemistry of the whole atmosphere. Phenomena like plastic and global warming are not only spatially large. They are temporally massive as well, not going anywhere for the foreseeable future. "Seven percent of global warming effects will still be manifest a hundred thousand years from now, slowly being absorbed by igneous rocks. That's more than ten times all of recorded history so far, a preposterously high number."[58]

In the Anthropocene, humans are facing a proliferation of objects that are massively large and long-lasting relative to the scale of a human life. These are hyperobjects—"things that are massively distributed in time and space relative to humans" (e.g., black holes, oil fields, the biosphere, Styrofoam, global capitalism, the Internet, etc.).[59] Humans encountered hyperobjects before the Anthropocene, but in the Anthropocene hyperobjects are proliferating wildly, as can be seen in economic globalization, telecommunications networks, global ecological crises like climate change and mass extinction, and scientific discoveries of massive cosmological objects, including extrasolar planets, supernovas, billions of galaxies, and the Cosmic Microwave Background—the radiating remains of the Big Bang. Though the Big Bang may seem like the distant past, its faint signal is everywhere, visible in the static of a television set.

On one hand, hyperobjects are horrifyingly close, imposing themselves on the intimacy of coexistential spatiality. They are already part of your life. Even if you deny climate change, you are inside of it. You could not deny it if you were not already haunted by its arrival. You are woven into the tapestries of cosmological and biological evolution. You are part of a mass extinction event the likes of which have not been seen for 65 million years. You are inside the global greenhouse of global capitalism, where even the most severely excluded or staunchly resistant person still lives with the impacts of the deregulation of markets. When we reflect on them, hyperobjects do not always seem that close, but they are there. Morton finds the warning etched into the passenger side mirror of American cars to be an appropriate slogan for an ontology oriented toward hyperobjects: "OBJECTS IN MIRROR ARE CLOSER THAN THEY APPEAR."[60]

On the other hand, hyperobjects are not only astonishingly close, "they are also uncanny."[61] As close as they are, as deep inside of you as your genome or as all-encompassing as the climate, hyperobjects are nonetheless ungraspable. You can never see the climate. You can see the weather change. You can see a drought ravaged landscape. You can contemplate statistical models of the climate. Yet none of those things is the climate. Grasping all life on Earth, it remains ungraspable. In close contact, it remains untouched, intact. It is everywhere, yet the singularity of its existence cannot be located anywhere. Global warming might exist for tens of thousands of years, yet the singularity of its existence is not present in any one of those years or in their sum. Hyperobjects can thus be described as "nonlocal and atemporal."[62] Never here or there, never now or then, hyperobjects are strange strangers whose arrival distorts human scales of space and time. Global warming is here on Earth, now, and it will be there in the future, yet it is also far weirder than that, as it compels humans to rethink what here and now means, to attune our spatiotemporal sensibilities to its arrival. It weighs on us like an imperative to change the space and time of our lives. Deborah Bird Rose makes this point regarding the unbearable emergence of a narrative out of the horror of increasing extinctions, a narrative that enjoins us "to rethink everything we thought we knew about who we are and how to live within the imperiled family of life on Earth."[63]

After humanism, anthropocosmology cultivates a different human, attuned to strange strangers in a mesh of ecological coexistence. While phenomenological attention, poetic imagination, and religious practices facilitate such attunement, all techniques for anthropocosmic ecstasy are dumbfounded by the arrival of hyperobjects in the Anthropocene. Those techniques are indispensable, yet they must retune their modes of attunement, resonating with the massively absurd tunes of hyperobjects. Three adjectives can characterize the detuned (*absurdus*) attunement of humans in the Anthropocene. Hyperobjects render humans hypocritical, weak, and lame.

The time of hyperobjects is a time of hypocrisy insofar as inside and outside have imploded, leaving no meta-position from which one can criticize or evaluate a hyperobject. We are all part of the problem. There is nowhere else to go. In the Anthropocene, "away" is another somewhere, very close by. Under the weight of hyperobjects, we are all hypo-critical, under (*hypo*) the weight of discernment and decision. Even the staunchest climate activist emits CO_2, which for a cynic negates the activist's message. The cynic, with a desire for moral purity and intolerance for compromise, fails to comprehend that an ecological emergency is already under way. There is no outside position from which to contemplate the spheres of planetary coexistences. The cynic fails to let the emergency emerge, an emergency in which all human are implicated and nobody is purely and simply innocent. In the Anthropocene, we are all hypocrites, but the cynic is a hypocrite who denies and criticizes hypocrisy, being hypocritical about hypocrisy.[64]

The cynic is a beautiful soul—one who sees only universality and harmony, turning a blind eye or a rolling eye to anything that compromises or contaminates the purity of those ideals. The "beautiful soul" is Hegel's term for a stage or moral-psychological development in which someone is "entangled in the contradiction between its pure self and the necessity of that self to externalize itself and change itself into an *actual* existence."[65] Unable to bear that contradiction, the beautiful soul retreats inward into a purity of spirit. "Beautiful soul syndrome," as Morton terms it, names the position of the cynic in the Anthropocene. "Beautiful me over here, corrupt world over there."[66]

> Such an attitude *directly is* the evil that it perceives over yonder, argues Hegel. Evil is the gaze that sees evil as a thing to be eradicated, a pathological symptom to be excised. This is the umbrella subject position of modernity. Whether one is a vegetarian or a fundamentalist, an end-times Christian or simply a flaneur, the world is seen predominantly as a product "over there," suspicious, seductive, slightly or terribly evil. The beautiful soul fails to acknowledge the ugliness that underwrites its seeming beauty. The disavowal of its own "evil," shifting and ambiguous perception, is the cause of what it sees over yonder in the world.[67]

By avoiding sincere engagements in coexistence, the beautiful soul moves through life as through a nihilistic escape route, attempting to find some clarity and purity away from the ambiguities of the coexistence. All engagements always result in some compromise, complicity, or unruly complexity. If you do not want to turn away cynically, you must embrace the irony of the situation. Turned inside-out and upside-down, being-in is being-under: being-sincerely-hypocritical.

The time of hyperobjects is a time of weakness because humans are no longer in a superior or transcendent position relative to other beings. Intimately intertwined in ecological coexistence, humans are vulnerable and exposed to other humans, nonhumans, and things that upset our distinctions between human and nonhuman, animate and inanimate, like Odradek—the spool of thread that weirdly moves, talks, and laughs. As with hypocrisy, weakness is not a bad thing. It is a bad thing for the robustly armored human of modern humanism, the skin-encapsulated ego walling itself off from intimacy. Weakness is the vulnerability of being with others. "Weakness determines the capacity for tuning."[68] It is the vulnerability of resonating with the detuned tunes of strange strangers. This vulnerability is the precondition for any connection. It is generative. Weakness is rich in humus, composting humanity in humility.

The time of hyperobjects is a time of lameness because every being is fragile and inconsistent, torn between what is and is not itself. We see rain but no climate, blog posts but no Internet, a dead bird but not mass extinction, a supermarket but not capitalism. As with hypocrisy and weakness, lameness is a feature of all beings in the Anthropocene. All objects are hypocrites insofar as their appearances are different from their realities. All objects are weak insofar as they are enmeshed with others. All objects are lame insofar as they fail to coincide with themselves totally.[69] If you are looking for the biosphere, when you look at a cat or any organisms you are looking at that sphere of life. Yet, at the same time, you are not looking at the biosphere. You are looking at that cat. The biosphere is and is not a cat; the cat is and is not the biosphere. Nothing is itself. Everything is lame. Again, this is not a bad thing unless you are attached to the hypertrophic subject of modern humanism or to any fantasies of an omnipotent power or ultimate ground that grants purity, strength, and consistency.

After humanism, the "hypersubjects" of modernity no longer make sense. Intimate coexistence in the Anthropocene calls for weak, lame, hypocritical subjects: "hyposubjects."

> Hypersubjects are typically, but not exclusively white, male, northern, well-nourished, and modern in all senses of the term. They wield reason and technology, . . . they seek transcendence; they get very high on their own supply of dominion. Do you want to know what is irritating hypersubjects today? The fact that hyperobjects are whispering in their ears, whispering that this being and time that they have fashioned in their image and for their own convenience is dying. The voices in their heads say that there is no time for hypersubjects any more. It is hyposubjectivity, rather than hypersubjectivity, that is becoming the companion of the hyperobjective era. . . . Hyposubjects are the native species of the Anthropocene and are only just now beginning to discover what they might be and become. . . . They turn things inside out and work miracles with scraps and remains. . . . If we wish to thrive, it is as hyposubjects that we will become human (again).[70]

To become anthropocosmic in the Anthropocene is to become human as a hyposubject—weak, lame, and sincerely hypocritical, recognizing oneself as a strange, vulnerable, fragile stranger intimately enmeshed with other strangers far beyond comprehension and control.

ECOLOGICAL HUMANITIES

Hyposubjects are abominable subjects, moving away from the human (*ab-homine*) to become human once again, a human cracked open by the inhuman, humbled by the weight of intimate alterities. Hyperobjects thus compel those who use poetic, phenomenological, and religious techniques for attuning to anthropocosmic resonances to become hypocritical, weak, and lame. Of course, poets, philosophers, and religious practitioners might already be hypocritical, but are they sincerely hypocritical? Sincere hypocrisy affirms the complicity and ambiguity of having your internal reality mixed up with the external world, such that you cannot judge anything from the outside. Anthropocosmic techniques are weak insofar as they remain exposed, vulnerable to the insistent demands of alterity. Anthropocosmology is lame insofar as an anthropocosmic experience fails to coincide with itself; it is an experience of my spine becoming a mountain, an experience of the emptiness of beings, an experience of the intimate immensity of Jupiter as I glance into the evening sky, but if I look for the anthropocosmos, it never shows up.

Hypocrisy, weakness, and lameness are not bad news unless you were hoping that anthropocosmology would be a humanism or posthumanism writ large, announcing the arrival of some kind of superhero superhuman or monstrous mutant. Still at the beginning, square one, we do not know who we are. Anthropocosmology does not happen in the position of one who knows, but in a disposition that is radically open to intimate alterities, which means more open to grieving, despair, and anxiety, more open to unheard-of uncanniness, and more open to considering the perspectives of others. This is all good news if you were hoping for a more radical role for the humanities in response to the conditions of ecological emergency. Just as poetic, philosophical, and religious techniques of anthropocosmic ecstasy are made more adaptable insofar as they are rendered more vulnerable, fragile, and inconsistent, so too is the interdisciplinary approach to ecological perspectives across fields like poetics, philosophy, and religious studies, in short, across the humanities.

The proliferation of hyperobjects is facilitating the emergence of ecological humanities, unchaining humanities from its fixation on the now defunct human of classical and modern humanisms, opening up to anthropocosmic correspondences and multispecies kinship. After God, after nature, after the hypersubjective human, the nature/culture and subject/object dualisms that support distinctions between the sciences and humanities no longer hold, at least not in the same way. "The ecological humanities," as Rose says, "aims to work across the great divides in knowledge that have enabled us to sustain a faulty image of humanity, an image that holds humans apart, and in control."[71] Some areas in which humanities scholars are working across those divides are ecocriticism, environmental justice, science and technology studies, multispecies ethnography, environmental ethics, and the study of religion and ecology. Working across those divides means letting something new emerge at the "abrasive edges between knowledge systems."[72] This means that the study of ecological perspectives in the humanities is not just for fun, when you want to curl up with a book about the animals in Buddhist scriptures or the waters of the poetic imagination. Inquiry in the humanities is not a hobby for the weekend while the sciences do the real work during the week. The humanities are not a candy coating on the medicine of science, technology, engineering, and mathematics. It is working alongside those other systems to facilitate the emergence of new possibilities for coexistence on Earth. A poet, a physicist, a sociologist, and a philosopher walk into a bar not because it is the beginning of a joke but because they are collaborating on the construction of different ways of thinking, feeling, and acting amid the unparalleled uncertainty and interconnectedness of the Anthropocene.

This is not to say that hyperobjects have rendered the humanities equal to the sciences in their capacity to articulate ecological phenomena. On one hand, the methods of the humanities and sciences are different and not comparable or measurable from any meta-position outside of them, which would only beg the question of the meta-position that measures the validity of that meta-position. On the other hand, if they are comparable, one could risk the hypothesis that the humanities have a more intimate role to play in articulating the specificities of ecological coexistence. That is for two reasons. First, the interdisciplinary field of ecological humanities "emerged specifically to address the fact that current ecological problems, including extinctions, climate change, toxic death zones, water degradation, and many others, are anthropogenic events."[73] In the Anthropocene, the modes of inquiry most intimately connected to *anthropos* are of crucial importance for understanding and responding to the anthropogenic transformations of Earth. Second, even aside from the anthropogenic cause of current ecological problems, the humanities is closely attuned to ecological phenomena because humanities scholars specialize in the kind of ambiguity, paradox, and uncertainty that is currently distorting human horizons through the arrival of hyperobjects. Technoscientific projects tend to minimize the risk associated with indeterminacy and ambiguity while humanities inquiries affirm human participation in an "ecology of risky attachments," remaining exposed to the exigencies of coexistence.[74] Disambiguation has its place; it would be important to abjure paradox in order to ensure, for instance, that airplanes are not simultaneously crashing and not crashing. Poets, monks, and philosophers can crash and not crash, finding coexistential truth in contradictions: the other appears by not appearing. In that sense, inquiry in the humanities is more equipped to talk about the reality of hyperobjects, not to measure or calculate hyperobjects, but to attune to their irreducibly singular reality.

Inquiries in ecological humanities take up literary, philosophical, religious, and cultural perspectives in order to address multiple facets of current ecological problems. It is important to bear in mind that one of those facets is the irreducible face of the real. In other words, the field of ecological humanities is ontological and not merely an exercise in critique, epistemology, or ethicopolitical engagements. To miss that ontological detail is to inscribe ecological humanities back into the usual dichotomies between values, subjects, meaning, and humans on one side and facts, objects, matter, and nonhumans on the other, or it pushes ecological humanities into the cynical position of the beautiful soul who does not participate in coexistence but sits back and criticizes everyone who falls for the usual dichotomies and power plays. If people get past the obstacle of thinking that the humanities are not relevant to ecology but are only weekend endeavors for recreation, the next obstacle to get past is thinking that humanities scholars only articulate different human ways of knowing, deciding, and making meaning, leaving ontology to the scientists or abandoning it altogether in critique. Even when scholars affirm the entanglement of knowing and being, the gravity of the great divides tends to pull them into the usual definitions, for instance, bringing religion and ecology together by putting scientific narratives on the side of ontology and religious narratives on the side of epistemology.[75]

Amid hyperobjects, it is glaringly apparent that no single perspective could possibly have a monopoly on what it means to be, and likewise, no area of inquiry is excluded from interpreting the truth of beings, not poetics, philosophy, physics, biology, economics, mathematics, not to mention the perspectives of chestnut trees, pangolins, sloths, mountains, cars, rivers, and dust. Weak, lame, and hypocritical, nothing is excluded from interpreting beings, and no interpretation can grasp the unfathomable singularity of any being. In any field of study, when you approach questions of ontology, you encounter the inexplicable strangeness of things, and you have to consider the interpretative and metaphorical dynamics of existence, which is to say, you have to consider the humanities. This does not mean that humanities scholars are now in charge. Humanities discourses could benefit from a lot more weakness, lameness, and sincerity, softening the impenetrable walls of highly specialized and gratuitously precise language games. "Humanities scholars have some very good and important ideas, if only they would let others read them."[76]

Becoming vulnerable, with ironic hesitation, humanities scholars could tend less toward the position of cynical know-it-alls and more toward the disposition of sincerely hypocritical contributions to planetary coexistence, supporting the efforts of humans to become human and tangle with hyperobjects amid ecological emergency. If this makes it sound like the field of ecological humanities provides a solution or an answer, I should apologize for not articulating this point more carefully. By supporting the efforts of humans to become human, anthropocosmic, enmeshed with intimate alterities, ecological humanities opens up more problems, not solutions, more compelling others insisting on justice. It lets the emergency emerge. Ecological humanities cannot make everything okay. Working across great divides—sciences and humanities, humans and nonhumans, nature and culture, religion and ecology—ecological humanities does not thereby make those oppositions fit together harmoniously. That is beautiful soul syndrome, seeing "only reconcilable and federative differences, far removed from bloody struggles. The beautiful soul says: we are different, but not opposed."[77]

The rifts between and within strange strangers are real, and so are the oppositions, struggles, and conflicts between humanities and sciences. "Humanities and sciences," as Morton observes, "hold broken pieces of a jigsaw puzzle, pieces that might not fit together."[78] The production of knowledge in ecological humanities is about getting involved in the struggles of coexistence, not about making everything fit. "Ecognostic jigsaws are never complete."[79] Following Blake, Morton is not going to fall for any scheme of "fitting & fitted."[80] If scholars of ecological humanities or practitioners of anthropocosmology aim to make everything fit, then they are turning their backs on the vicissitudes of coexistence. If they are tangling with hyperobjects, weakening the humanities in anthropocosmic compost, then, perhaps, out of all of this hypocrisy, lameness, and weakness there opens up our chance of becoming human, once again and still for the first time.

NOTES

1. Paul Ehrlich, *The Population Bomb*, rev. ed. (Jackson Heights: Rivercity Press, 1975).
2. Murray Bookchin, "Social Ecology versus Deep Ecology," in *Environmental Ethics: Readings in Theory and Application*, 5th ed., ed. Louis P. Pojman and Paul Pojman (Belmont: Thomson Wadsworth, 2008), 245.
3. Naomi Klein, *The Shock Doctrine: The Rise of Disaster Capitalism* (New York: Picador, 2007).
4. Donna Haraway, "Anthropocene, Capitalocene, Plantationocene, Chthulucene: Making Kin," *Environmental Humanities* 6 (2015), 162.
5. Kierkegaard, *Papers and Journals*, 32.
6. Sartre, *Existentialism Is a Humanism*, 20.
7. Martin Heidegger, "Letter on 'Humanism,'" in *Pathmarks*, ed. William McNeill, trans. Frank A. Capuzzi (Cambridge: Cambridge University Press, 1998), 241.
8. Ibid., 248–50, 266. For Heidegger, essence and existence are not so neatly separated. *"The 'essence' of Dasein lies in its existence."* Heidegger, *Being and Time*, 67.
9. Maurice Merleau-Ponty, *Phenomenology of Perception*, trans. Colin Smith (New York: Routledge and Kegan Paul LTD, 1998), xi, xiv. Merleau-Ponty notes that his concept of the world is present in Heidegger's *Being and Time* and also in Husserl's articulation of the "natural world-concept" (*natürlicher Weltbegriff*) and the "life-world" (*Lebenswelt*), the latter of which became the central theme of Husserl's later works (ibid., vii). See Edmund Husserl,*The Crisis of European Sciences and Transcendental Phenomenology: An Introduction to Phenomenological Philosophy*, trans. David Carr (Evanston: Northwestern University Press, 1970), 121–43.
10. Ibid., 141.
11. Maurice Merleau-Ponty, *The Visible and the Invisible*, trans. Alphonso Lingis (Evanston: Northwestern University Press, 1968), 38.
12. Ibid., 139–40.
13. Ibid.,146–47, 248–50.
14. Ibid., 248.
15. According to Ted Toadvine, David Abram's ecophenomenology tends to employ Merleau-Ponty's philosophy in this way, marginalizing reflection and mediated experiences in favor of immediacy, and occluding alterity with continuity. Abram disagrees with that assessment. Ted Toadvine, "Limits of the Flesh: The Role of Reflection in David Abram's Ecophenomenology," *Environmental Ethics* 27.2 (2005). David Abram, "Between the Body and the Breathing Earth: A Reply to Ted Toadvine," *Environmental Ethics* 27.2 (2005). It should be noted that, although Abram disagreed with Toadvine in 2005, Abram's more recent monograph—*Becoming Animal: An Earthly Cosmology* (New York: Pantheon Books, 2010)—still admits of the problems Toadvine diagnoses, tending to privilege immediacy and continuity over mediation, reflection, and alterity. See my review of this book, Sam Mickey, "Review of *Becoming Animal: An*

Earthly Cosmology, by David Abram," *Worldviews: Global Religions, Culture, and Ecology* 16.1 (2012).

16. Maurice Merleau-Ponty, *Signs*, trans. Richard C. McCleary (Evanston: Northwestern University Press, 1964), 39.

17. Merleau-Ponty, *The Visible and the Invisible*, 147.

18. Ibid., 146.

19. This extension of intentionality and meaning to nonhumans is particularly evident in Merleau-Ponty's lectures on the concept of nature. Merleau-Ponty, *Nature: Course Notes from the Collège de France*, trans. Robert Vallier (Evanston: Northwestern University Press, 2003). For a detailed look at this aspect of Merleau-Ponty's thought, see William S. Hamrick and Jan Van der Veken, *Nature and Logos: A Whiteheadian Key to Merleau-Ponty's Fundamental Thought* (Albany: SUNY Press, 2011).

20. Suzanne L. Cataldi and William S. Hamrick, eds., *Merleau-Ponty and Environmental Philosophy: Dwelling on the Landscapes of Thought* (Albany: SUNY Press, 2007).

21. Merleau-Ponty, *Signs*, 110.

22. Merleau-Ponty figures into Andy Fisher's "ecological existentialism." Fisher, *Radical Ecopsychology*, 267n9.

23. Bachelard, *The Poetics of Space*, 47.

24. Gaston Bachelard, *The Flame of a Candle*, trans. Joni Caldwell (Dallas: Dallas Institute Publications, 1988), 2.

25. Bachelard, *The Poetics of Space*, 171.

26. Ibid., 203.

27. Ibid., 155.

28. Ibid., 183–85.

29. Ibid., 22.

30. I outline the genealogy of Bachelard's and Eliade's uses of anthropocosmic terminology elsewhere, Sam Mickey, "Contributions to Anthropocosmic Environmental Ethics," *Worldviews: Environment, Culture, Religion* 11.2 (2007).

31. Mircea Eliade, *Images and Symbols: Studies in Religious Symbolism*, trans. Philip Mairet (Princeton: Princeton University Press, 1991), 36.

32. Mircea Eliade, *The Sacred and the Profane: The Nature of Religion*, trans. Willard R. Trask (San Diego: Harcourt Brace Jovanovich, 1987), 169.

33. Ibid., 169–70.

34. Ibid., 8–16.

35. Rudolf Otto, *The Idea of the Holy*, trans. John W. Harvey (London: Oxford University Press, 1958), 5–32.

36. Eihei Dōgen, *Moon in a Dewdrop: Writings of Zen Master Dōgen*, ed. Kazuaki Tanahashi (New York: North Point Press, 1985), 70.

37. Ibid., 164.

38. Hermes Trismegistus, "*The Emerald Tablet* (*Tabula Smaragdina*)," in *The Alchemy Reader: From Hermes Trismegistus to Isaac Newton*, ed. Stanton J. Linden (Cambridge: Cambridge University Press, 2003), 28.

39. Tu Weiming, "The Ecological Turn in New Confucian Humanism: Implications for China and the World," *Daedalus* 130.4 (2001): 243.

40. This is a phrase commonly used by Alan Watts, a prolific icon of countercultural philosophy. Alan Watts, *Eastern Wisdom, Modern Life: Collected Talks, 1960–1969* (Novato: New World Library, 2011), 75–77.

41. Slavoj Žižek, *In Defense of Lost Causes* (London: Verso, 2008), 140.

42. John D. Caputo, *More Radical Hermeneutics: On Not Knowing Who We Are* (Bloomington: Indiana University Press, 2000).

43. Morton, *The Ecological Thought*, 8.

44. Another way to translate this phrase (*tout autre est tout autre*) that is closer to rendering its palindromic character would be "*every other (one) is every (bit) other.*" Jacques Derrida, *Politics of Friendship*, trans. George Collins (London: Verso, 1997), 232.

45. Caputo, *More Radical Hermeneutics*, 251.

46. Bachelard, *The Poetics of Space*, 184.

47. Haraway says this in an interview. Nicholas Gane and Donna Haraway, "When We Have Never Been Human, What Is To Be Done?," *Theory, Culture and Society* 23 (2006): 140.

48. Donna Haraway, "SF: Science Fiction, Speculative Fabulation, String Figures, So Far." Pilgrim Award Acceptance Talk Presented at the Science Fiction Research Association (SFRA) Conference, Lublin, Poland (July 2011), 10.

49. Morton, *The Ecological Thought*, 113.

50. Ibid.

51. Derrida, *Politics of Friendship*, 232.

52. Jacques Derrida, *Specters of Marx: The State of the Debt, the Work of Mourning, and the New International*, trans. Peggy Kamuf (New York: Routledge, 1994), 196.

53. Morton, *The Ecological Thought*, 143n72.

54. Alphonso Lingis, *The Imperative* (Bloomington: Indiana University Press, 1998), 68–69.

55. Ibid., 63.

56. For more on the possibility that the Anthropocene could be a boundary event (a transitional period) and not a full epoch, as well as more regarding the contested naming of this event/epoch, see Haraway, "Anthropocene, Capitalocene, Plantationocene, Chthulucene," 160. See also, Donna Haraway et al., "Anthropologists are Talking—About the Anthropocene," *Ethnos* (2015).

57. Morton, *Dark Ecology*, 24.

58. Morton, *Realist Magic*, 57.

59. Morton, *Hyperobjects*, 1.

60. Ibid., 27.

61. Ibid., 28.

62. Ibid., 47.

63. Rose, *Wild Dog Dreaming*, 146.

64. Morton, *Hyperobjects*, 148.

65. Georg Wilhelm Friedrich Hegel, *Hegel's Phenomenology of Spirit*, trans. A. V. Miller (Oxford: Oxford University Press, 1977), 406.

66. Morton, *Hyperobjects*, 154.

67. Timothy Morton, "X-Ray," in *Prismatic Ecology: Ecotheory Beyond Green*, ed. Jeffrey Jerome Cohen (Minneapolis: University of Minnesota Press, 2013), 316.

68. Morton, *Hyperobjects*, 176.

69. Ibid., 195.

70. Dominic Boyer and Timothy Morton, "Hyposubjects," *Theorizing the Contemporary, Cultural Anthropology* (January 21, 2016).

71. Deborah Bird Rose, "The Ecological Humanities," in *Manifesto for Living in the Anthropocene*, ed. Katherine Gibson, Deborah Bird Rose, and Ruth Fincher (Brooklyn: Punctum Books, 2015), 3.

72. Ibid.

73. Ibid., 1.

74. Lesley Instone, "Risking Attachment in the Anthropocene," in *Manifesto for Living in the Anthropocene*, ed. Katherine Gibson, Deborah Bird Rose, and Ruth Fincher (Brooklyn: Punctum Books, 2015), 36.

75. For instance, in Bauman's viable agnostic approach to religion and ecology, he affirms the intertwining of knowing and being, which is emphasized in his use of the portmanteau "epistem-ontology," yet he nonetheless frames his telling of a "scientific story" as "roughly a new ontology" and his telling of a "religious story" as "roughly a new epistemology." Bauman, *Religion and Ecology*, 5, 201.

76. Morton, *The Ecological Thought*, 13.

77. Deleuze, *Difference and Repetition*, 52.

78. Morton, *The Ecological Thought*, 12.

79. Morton, *Dark Ecology*, 8.

80. Morton, *The Ecological Thought*, 12. Morton is referring to Blake's comments in the margins of a passage from William Wordsworth. Responding to Wordsworth's saying, "The external World is fitted to the Mind," Blake says. "You shall not bring me down to believe such fitting & fitted." Blake, *The Complete Prose and Poetry of William Blake*, 667.

EIGHT

Looking Good

After humanism, after God and nature, humans can become humans in all of their effulgent differences, unchained from the untenable habits, norms, conventions, and stereotypes orienting humanism, modernism, and consumerism around "the hegemony of the homogeneous."[1] Ecological emergency is severely weakening the hypertrophic subjects who seek power over others. It is opening the way for different humans, hyposubjects, who are far stranger and with much greater humility, as well as greater irony. Hypersubjects try to make everything and everyone look the same, like Descartes' thinking thing (*res cogitans*), Kant's good will, the beautiful souls of Romantics, or one of Coke's consumers. Hyposubjects affirm coexistence under the weight of intimate alterities, exposed to the hypocrisy, weakness, and lameness of being human in the Anthropocene. Hyposubjects diffract along multiple lines of difference. "Hyposubjects are necessarily feminist, colorful, queer, ecological, transhuman, and intrahuman."[2] Those differences can be understood along countless interpretive lines. I would like to suggest that they can be interpreted as differences of looking good or, in other words, differences of artistic body transformation, where artistic transformation is not about fanciful decoration or spectacle but is, as philosophers like Foucault understand, about living life in excess of given requirements or disciplines, even in excess of aesthetic standards.[3] Elizabeth Grosz elaborates on the practice of living life as a work of art.

> This is not only to act different, but also to be acted on differently, i.e., to sense differently, not only developing each of the senses beyond its usual biological reach but to inquire into the limits and transformability of biology itself. It is also to synaesthetically cross-map the senses onto each other, to experiment with how to sense differently, how to use senses in terms of the range and scope of the other senses, to ex-

> plore how each sense functions or is capable of functioning quite differently from its assumed and normalized role. To develop these alternatives is inherently invested in artistic activity.[4]

In short, to develop these alternatives is about "looking good in both senses: to look at something well (outside the domain or order of the gaze), and to be looked at as well."[5] Looking good orients itself toward and for the other, looking at others and being looked at by others. This is the aesthetics of intimate alterity.

Looking good is about showing off your strangeness, getting weird. It is not done for survival, or at least not exclusively so. The ornate coloring and plumage of birds-of-paradise put those birds' survival at risk, exposing them to predators. Consider the sexual habits of cephalopods. Dumpling squid have sex for approximately three hours at a time, and they have sex frequently throughout their yearlong life.[6] The costs of this extensive sex life are quite high. Three-hour mating encounters are exhausting, as indicated by the slow swimming of squid after sex. At that point, it is difficult to perform any survival tasks, such as finding food, avoiding predators, and finding more mates. Sex is not oriented exclusively toward survival, especially not a survival of those who are fit or fittest. Quite to the contrary, sexual cephalopods involve exorbitant expression that, like Blake, abjures the "fitting & fitted."[7] Of course, the reproductive part of the sexual encounter can be understood in terms of survival, but it could be the sexiest and not necessarily the fittest squid that passes on its genes, and insofar as sexiness can counterbalance survivability and fitness, one must recognize that survival is not the exclusive or necessarily primary operation.

In any sexual encounter, there is a rift between survival values of death and birth and the exorbitant values of attractiveness and appeal—the values of looking good. In terms of Darwin's theory of evolution, this is the rift between natural selection, which involves functions of birth and death, and sexual selection, which involves functions of appeal and attraction that can counteract the survival functions of natural selection.[8] Natural selection involves the struggle of all members of a species to exist, whereas sexual selection arises with the struggle of individuals to stand out from other members of their species. Moreover, contrary to sociological interpretations of evolution, which typically relegate females to passive roles, sexual selection redistributes agency across the sexes, recognizing active roles of females.

> Sexual selection depends on the success of certain individuals over others of the same sex in relation to the propagation of the species; while natural selection depends on the success of both sexes, at all ages, in relation to the general conditions of life. The sexual struggle is of two kinds; in the one it is between the individuals of the same sex, generally the male sex, in order to drive away or kill their rivals, the females

> remaining passive; whilst in the other, the struggle is likewise between the individuals of the same sex, in order to excite or charm those of the opposite sex, generally the females, which no longer remain passive, but select the most agreeable partners.[9]

A female might choose to mate with a male bird that exhibits an appealing song and plumage, even if those exuberant sounds and colors diminish survival value. When mating does occur, reproduction can be one of its outcomes but need not be. Along these lines, against the fitness craze of Social Darwinism and sociobiology, Darwin's own writings on evolution appear useful for understanding practices of looking good, specifically for situating human practices of looking good within ecologies of exorbitantly expressive nonhumans.

As Grosz observes, Darwin's concept of sexual selection is resonant with some aspects of feminist theories of sexual difference, specifically that of Luce Irigaray.[10] Putting Darwin's theory of sexual selection in dialogue with Irigaray's sense of sexual difference provides a context for rethinking the existentialist struggle to live life like a work of art, broadening the scope of that struggle to include a spectrum of differences that far exceeds the default white, heterosexual, male, human subject position of existentialists. This spectral expansion of existentialism through a feminism of difference is fundamental to coexistentialism. For instance, Morton refers to Irigaray as an example of one of the "prominent recent theorists of 'coexistentialism.'"[11] Irigaray's coexistentialism is evident in her articulation of the irreducible alterity and dissent pervading coexistence. "The rift between the other and me is irreducible. To be sure we can build bridges, join our energies, feast and celebrate encounters, but the union is never definitive, on pain of no longer existing. Union implies returning into oneself, moving away, dissenting, separating."[12] Similarly, Rose finds feminist theories of difference to be crucial partners in her "ecological existentialism," particularly in light of the works of Donna Haraway and Val Plumwood.[13] Rose draws most frequently from Plumwood, her friend and colleague. Like Irigaray, Plumwood affirms the mesh of connectedness as well as the strangeness and dissent of alterity. "We need to understand and affirm both otherness and our community in the earth."[14] In other words, we need to understand and affirm "a complex, interacting pattern of both continuity and difference."[15]

With a sufficiently radical theory of sexual difference, which can integrate Darwin's insights regarding sexual selection, differences of gender, race, and species can be understood and affirmed as different ways of looking good, acting differently and being acted on differently through exorbitantly sexy, colorful, ecological body transformation. To be sufficiently radical as to include human and nonhuman beings, sexual difference must be understood ontologically and not merely biologically or culturally. After tracing an ontological conception of sexual difference and its relevance to the task of looking good as and for the feminist, queer, prismatic, ecological hyposubjects of the Anthropocene, I consider sexual difference at work in a perhaps unlikely example: psychedelic experiences.

CHANGING SEXUAL DIFFERENCE

On the fashionable and relatively trendy side of philosophical and literary movements, existentialism has garnered a notably sexy reputation, as is obvious from a cursory glance online searching for the portmanteau, "sexistentialism." However, for the most part, the predominant figures in existentialism tend not to understand and affirm sexual difference. Those figures tend to be men, and those men often (but not always) portray the kind of sexist assumptions typical in their cultural milieus, relegating women to passive or emotional caregiver dispositions subordinate to the more active or rational will that characterizes the subject positions of men. Kierkegaard provides a good example of this in *The Sickness unto Death*, where his pseudonymous persona Anti-Climacus, mistaking patriarchal norms for essential nature, remarks that "woman's nature is devotion (Hengivenhed), submission (Hengivelse), and it is unwomanly if it is not so."[16] Men can be devoted, but it is not their nature. Kierkegaard goes on to say that, whereas a man is himself, has a self, women lose themselves in devotion, and this lost self is the only self for their gender. Anything else would be unwomanly. A woman loses herself to a man, "and only thus is she happy, only thus is she herself."

At best, Kierkegaard's writings are ambiguous on topics of sexual difference, opening some possibilities for understanding and affirming the actual existence of women, and at worst, Kierkegaard's contributions are simply sexist and misogynistic.[17] The same can be said for Nietzsche and Heidegger, whose legacies are likewise fraught with ambiguous perspectives on women and sexual difference, yet those two thinkers nonetheless provide resources that have been taken up by numerous feminist philosophers.[18] Simone de Beauvoir stands out as someone—a woman—for whom existentialism is a feminism. Existence is not determined in advance by any essence or destiny, such that who a woman is depends on how she chooses to live in response to her conditions. This is liberating insofar as it allows an actually existing individual to take responsibility for her existence, deciding for herself how she wants to live her life. Who she is depends upon who she becomes.

"One is not born, but rather becomes, woman. No biological, psychic, or economic destiny defines the figure that the human female takes on in society."[19] What defines the human female is the social mediation of civilization as a whole, which historically has defined the human male as the subject and the female as "*Other*" in the sense of an object lacking subjectivity, a castrated subjected—"the eunuch that is called feminine."[20] However, although de Beauvoir's analysis of the historical exclusion and subordination of female subjectivity is of inestimable value to feminism and philosophy, it is important to note that her analysis too has its own limitations for affirming sexual difference. Her dialectical (Hegelian) understanding of the constitution of subjectivity through the exclusion of an objectified other emphasizes separation at the cost of continuity. The vulnerability of being-with loses its intimacy and becomes entirely oppositional. For de Beauvoir, "a fundamental hostility to any other consciousness is found in consciousness itself; the subject posits itself only in opposition; it asserts itself as the essential and sets up the other as inessential, as the object."[21]

De Beauvoir's existentialism does not overcome the dualistic hierarchy marking the historical subjugation of women, the subject/object hierarchy. It simply makes it a genetic instead of static hierarchy. As Catherine Keller observes, this analysis "replaces the traditional substantial self, which simply and ever *is*, with a dynamic consciousness that *has become*."[22] It is laudable to change the sense of a subject enduring consistently and identically, replacing it with a subject who becomes, as that allows those who currently are not recognized as *being* subjects to nonetheless *become* subjects. However, de Beauvoir's existentialist feminism is a humanism, wherein the self is a hypertrophic subject—a transcendental ego that excludes and objectifies the other. "Transcendence then becomes a function of ontological hostility."[23] This hostility occludes the intimacy whereby the other resonates within the subject, the microspheric intimacy constituting the subject or the ethical intimacy compelling the subject toward responsibility. Becoming woman in this context is a matter of women participating in the hyperactive subjectivity of humanism. It is not a matter of bursting the bubble of skin-encapsulated egos and engaging in the struggle for coexistence.

Feminists of coexistentialism like Irigaray and Plumwood overcome the binary opposition of the transcendental subject and objectified other.[24] Accordingly, Irigaray supplements de Beauvoir's dictum, rephrasing it in a way that entangles being (transcendent, culture) and becoming (immanent, nature). "It's not as Simone de Beauvoir said: one is not born, but rather becomes, a woman (through culture), but rather: I am born a woman, but I must still become this woman that I am by nature."[25] Similarly, responding to Nietzsche's dictum, which enjoins you to become who you are, Irigaray again rephrases the being/becoming dualism in a way that circumvents the settled stability of the transcendental subject.[26] "Be what you are becoming, without clinging to what you might have been, what you might yet be. Never settle."[27] Irigaray exceeds the unsettling acrobatics of existentialism by deepening the commitment to never settle, to always fly farther into the strange tangles of coexistence. The style of existentialism is thus refashioned in her feminist writing.

> This "style," or "writing," of women tends to put the torch to fetish words, proper terms, well-constructed forms. . . . It is always *fluid,* without neglecting the characteristics of fluids that are difficult to idealize: those rubbings between two infinitely near neighbors that create a dynamics. Its "style" resists and explodes every firmly established form, figure, idea or concept. Which does not mean that it lacks style, as we might be led to believe by a discursivity that cannot conceive of it. But its "style" cannot be upheld as a thesis, cannot be the object of a position.[28]

It would thus be inappropriate to Irigaray's style to confine her fluid thinking into a thesis, postulate, or position. Her feminism of difference is a changing feminism with a changing difference.

Irigaray's writings contribute much toward psychoanalytic and phenomenological theories of subjectivity, but her work is more than a theory of the subject, as it also opens onto elemental and even ontological dimensions. Indeed, as Catherine Malabou observes, she is "one of the only people to think explicitly together ontology and gender differences."[29] Reinterpreting the subject in light of sexual difference requires a reinterpretation of everything. Irigaray strikes an anthropocosmic tone: "A revolution in thought and ethics is needed if the work of sexual difference is to take place. We need to reinterpret everything concerning the relations between the subject and discourse, the subject and the world, the subject and the cosmos, the microcosmic and the macrocosmic."[30] In this sense, Irigaray's coexistentialist feminism is not a humanism. In short, it is not anthropocentric but anthropocosmic. It burst the humanistic bubble and opens the human up to a more intimate exposure to the world of nonhumans. "Respect for the transcendence of the other as other within sexual difference will also allow us to relearn to *respect the cosmos as having an existence other than ours* and that is not at the disposal of our needs and our desires."[31] All modes of alterity can thus be understood in terms of the alterity of sexual difference. It is not only the difference between male and female humans or the differences of sexual dimorphic organisms or sexual reproduction. The whole of the natural world is opened up by this alterity which is not one, which means exactly that there is no whole of the natural world. "No one nature can claim to correspond to the whole of the natural. There is no 'Nature' as a singular entity."[32]

After nature, you could speak of being in terms of a singular plurality or plural singularity, but the point is that the singular is never single, and the many is never a plurality of singles. The basic unit of sexual difference is the couple. Not unlike the bipolar intimacy constitutive of Sloterdijk's bubbles (e.g., mother-child, child-placenta, face-to-face, and heart-to-heart), sexual difference refutes loneliness by beginning with two. A theory of sexual difference thus contributes not only to the political liberation of women from the violence wrought by patriarchal and androcentric modes of existence but also to an ontology that accounts for a radical alterity which is not one.

> Replacing the one by the two of sexual difference thus constitutes a decisive philosophical and political gesture, one which gives up a singular or plural being [l'être un ou pluriel] in order to become a dual being [l'être deux]. This is the necessary foundation for a new ontology, a new ethics, and a new politics, in which the other is recognized as other and not as the same: bigger or smaller than I, or at best, my equal.[33]

As Grosz points out, some interpreters of Irigaray fail to account for this ontological dimension of sexual difference, reading Irigaray only in terms of her psychoanalytic, ethical, and political perspectives on female humans.[34] That failure is evident in some common criticisms of Irigaray's work.

Critics accuse Irigaray of "privileging sexual difference over all other types of difference."[35] Irigaray's emphasis on the couple sounds heteronormative if you fail to account for the couple as the ontological structure of bipolar intimacy and, instead, reduce the couple to an anthropological or biological object. Similarly, her emphasis on sexual difference sounds racist and classist if the primacy of the couple means that the problems of oppression related to differences between men and women are privileged over the problems of oppression predicated upon racial and class differences. Charges of heterosexism, racism, and classism involve the incorrect assumption that Irigaray is emphasizing one kind of oppression at the exclusion of others. However, Irigaray never advocates focusing on any single form of oppression universally. She advocates women acting according to their own specific interests and needs, focusing on "the most immediately unbearable of her modes of oppression."[36] What is most immediately unbearable to you? It could involve nationality, occupation, class, race, age, gender, ability, religion, etc. If the oppression of the LGBTQ community is the most pressing problem, that should take precedence for you. If it is race and class where you experience oppression most unbearably, that should take precedence for you.

Rather than taking anything away from struggles against violence and oppression, sexual difference enhances their specificity, enhancing their ability to look good—to discern the detailed texture of differences constituting specific struggles. It is not a matter of privileging sexual difference over other struggles as much as it is a matter of pointing out "that how one experiences one's race, class, ethnicity, or religion is sexually differentiated."[37] For instance, immigration is experienced differently by men and women, and while those differences are entangled with many lines of difference (e.g., race, class, and ethnicity) that must be carefully tended, the specificity of someone's body must not be effaced if that person is to be treated respectfully and justly. This is not to say that sexual difference causes or produces other differences. For instance, poverty is experienced differently by men and women, and poor people can look good (intensifying one's capacity to act and be acted on differently); however, the cause of class differences among humans is not sexual difference. Sexual selection does not create poverty, socioeconomic systems and decisions do.

In the case of racial difference, in contrast to socially constructed differences like class and religion, sexual differentiation does seem to play a generative role. Contrary to colorblind nihilists, racial differences are not merely social constructs. They are living phenotypes, but racial differences are not thereby reducible to natural selection or any natural hierarchy of fitness, such as racist Social Darwinists would assume. "Darwin's claim, in brief, is that slight variations in skin color, facial features, or other bodily or psychical characteristics that were only simply individual variations have proved to be pleasing and attractive, resulting in sexual selection."[38] Gradually, over numerous generations, individual variations that appeal to locally specific standards of taste, attractiveness, and beauty form typological patterns of looking good, which contemporary discourse refers to with the term "race." Darwin considers the possibility that it is the beauty of skin color that led to the loss of hair in the transition from primates to hominids. The relative hairlessness of humans could be, as Grosz suggests, "a consequence of a desire to see the color of the skin, a desire to observe and be pleased by skin color, the most primary and obvious marker of race in our culture."[39]

An affirmation of sexual difference is thus also an affirmation of racial differences unchained from a single thing called "nature." In an affirmation of exorbitant body transformations, the colors of strange strangers are not assimilated into racist hierarchies that would privilege whiteness, and neither are they assimilated into environmentalist hierarchies which obsess over the greening of everything. "What is required, contrary to mainstream environmentalism—a term that is curiously close to *sexism* and *racism*—is," as Morton suggests, "ecology without nature."[40] Beyond white supremacy, green supremacy, or any supremacy, strange strangers find aesthetic affirmation in the context of a colorful ecology without nature, a "prismatic ecology."[41] Welding the prism of aesthetic differences to an essential nature is what racism is all about. "The racist is the one who fills in the gap between how a human appears and what a human is with some kind of metaphysical paste."[42] The racist is not looking good, not sensing differently, not sensing the strangeness of appearances but reducing appearances to some assumed, normalized, naturalized role.

You might still object that, while sexual difference does not privilege issues of sex and gender over other modes of oppression, insofar as it insists on the unique difference between male and female, sexual difference is still unjustly cisgendered and heteronormative. However, in Irigaray's ontology of intimate couples, sexual differentiation must be thought of not biologically or culturally but, like Sloterdijk's spheres, morphologically. Affirming the morphological specificity of male-female difference is about affirming the specificity of the body, without which there is no looking good—no affirmation of the exorbitant transformations of the body along diverse lines of sex, gender, reproduction, and family arrangements.

> Irigaray argues that whatever else one might be—whatever race, class, sexuality, nationality, ethnicity, and religion one might be assigned to—one is assigned as male, or as female, or in the mode of some identification with male or female. She questions, not homosexuality, nor ethnic identification, but only the disavowal of one's own morphological specificity. However queer, transgendered, and ethnically identified one might be, one comes from a man and a woman, and one remains a man or a woman, even in the case of gender-reassignment or the chemical and surgical transformation of one sex into the appearance of another. . . . The sexually specific body that is altered remains a sexually specific, if altered, body. Sexual difference has no one location, no one organ or condition.[43]

In short, the accusations of heterosexism, classism, and racism seem misguided, failing to recognize the ontological dimension of sexual difference, which opens up possibilities for queer, feminist, and prismatic modes of coexistence. Those accusations provide an indication that Irigaray's ontology seems relatively easy to miss, perhaps because she is somewhat ambiguous regarding the relationship between the ontological and sexual aspects of difference, sometimes collapsing the former into the latter.

It has been among a new generation of feminist theorists of difference that the ontological character of sexual difference has been more explicitly elaborated, specifically in the works of Grosz and other feminists associated with an emerging paradigm of feminist materialisms or materialist feminisms, what Grosz has called "corporeal feminism," which is not a reductionistic materialism but one that affirms the agency, meaning, and creativity of different bodies.[44] To further clarify the ontological dimensions of Irigaray's conception of sexual difference, the materialism of Catherine Malabou is particularly relevant. Engaging "deconstruction in a *new materialism,*" Malabou's materialism is oriented around a deconstructive conception of the form of matter.[45] Abjuring form/matter dualisms, Malabou defines materiality in terms of plasticity, which refers to three capacities simultaneously: to give form, receive form, and change form (transform). Those three capacities are exemplified by the plasticity of sculpture (giving), clay (receiving) and plastic explosives (transforming).[46] In *Changing Difference: The Feminine and the Question of Philosophy,* Malabou draws on her concept of plasticity and her experience and identity as a woman philosopher to develop a different conception of sexual difference. In short, sexuality is characterized by plasticity. Different identities and performances of gender are developments of plasticity. "If sex were not plastic, there would be no gender."[47]

Malabou uses plasticity to account for the difference between ontological difference and sexual difference. She follows Irigaray's definition of the feminine as "the fold of the lips to one another, a withdrawal that is so easy to force open, to breach, to deflower, but which at the same time also marks the territory of the inviolable."[48] This definition provides an ontological meaning that prevents the difference of the feminine, which can be thought as the difference constituting intimate alterity or hospitality to the other, from being reduced to the biological or cultural differences of women. However, she finds an ambiguity in Irigaray, one she also finds in Levinas, wherein the ontological difference of the feminine is sometimes confused with the sexual difference of women, such that the definition of the feminine in terms of folding lips, withdrawn yet disclosed, becomes a literalized relationship between the lips of a woman in binary opposition to the phallus.[49] Such literalization would mean that an ontology that situates itself in relationship to sexual difference would indeed perpetuate heteronormative binaries, and it would privilege struggles of gender and sexuality over struggles regrading race and class. To avoid such literalization, one must maintain a clear distinction between the ontological difference of the feminine from the sociocultural and biological differences of women.

A crucial problem here is that any ontological categorization of the feminine risks confining the feminine and actually existing women into a fixed essence, thus perpetuating the multiple modes of violence (sexual, symbolic, socioeconomic, cultural) against women in patriarchal and androcentric contexts. However, the antidote to essentialism cannot be the anti-essentialism found throughout deconstruction, gender studies, and queer theory, as that perpetuates a nihilistic violence that relativizes difference and fails to acknowledge the specificity reality of women or femininity at all. Moreover, this "anti-essentialist violence can play into the hands of ordinary violence against women in both the domestic and social realm."[50] Of course, the violence of a deconstructionist is not the same kind of material terror inflicted in social and domestic violence. "But even if it is only to reject it, we cannot avoid questioning the complicity between a domestic and social violence that refuses to give women a place and a theoretical violence that refuses to give women an essence."[51]

Rejecting the violence of essentialism, Malabou is after essence, following the deconstruction of essence. She does not remain with anti-essentialism but proceeds toward "a post-deconstructive meaning" of "what remains of the feminine after its deconstruction."[52] Speaking of remains is not speaking about debris or trash. "It refers not to a residue but to a kernel of resistance, a kernel with the strength of a new beginning, a live instance that still burns."[53] What remains after the feminine? What remains is a two-fold exposure to violence, a "schizoid pressure" of the social and the domestic, public and private, "inside and outside the home"; the specificity of that "overexposure to dual exploitation" is the paradoxically *inviolable* kernel of the feminine.[54] Malabou understands the harsh reality of violence against women, remarking that "no doubt woman will never become impenetrable, inviolable. That's why it is necessary to imagine the possibility of woman starting from the structural impossibility she experiences of not being violated, in herself and outside, everywhere."[55] This is a "minimal concept," a minimalist essentialism, avoiding the relativism of anti-essentialism without falling back on an essence that confines women within a fixed identity or definitive set of qualities.

After essentialism and after deconstruction, "woman" is empty. "Woman is nothing any more."[56] Yet being nothing is not nothing. It is exactly something, irremediably itself, and nothing more. What remains is the swarm of intimate alterities, just the swarm, just the dust of this planet.[57] The essence of woman is an essence without essence, a "resistant essence," "a *stamp of impossibility*," an irreducible specificity that folds together in a way that withdraws inviolably while remaining exposed to the violence by which essences are imposed or emptied.[58] In terms of Morton's "*object-oriented feminist ecology*," this essentialism of impossibility is "*weird essentialism*."[59] The essence of something is not a fixed identity or list of qualities but the fragile inconsistency whereby something is and is not itself, engaged in the aesthetic performance of exchanging its irreducible and inviolable specificity with its contingent and partial appearances exposed to others. A crucial difference between Morton and Malabou is that the latter seeks a new materialism, whereas Morton trains his focus on things, not matter and bodies, which risk effacing the specificity of actually existing things (individuals or collectives) by explaining them in terms of a substratum.[60] In any case, Morton and Malabou converge in recovering a feminine essence in terms of inviolable withdrawal, a stamp of impossibility. The impossible explodes with possibility.

This resistant essence of woman is found where there is no essence, in the plasticity of their bodies, giving, receiving, and transforming the woman who it is. According to Malabou, "the plasticity of difference" means that beings (women) and Being (the feminine) "exchange modes of being," which is to say, they "change from one into the other."[61] Being is being different, and being different means exchanging essence and manifestation. It means being changed by your essence while changing your essence, changing yourself as your essence. Women can thus become sites of the manifestation of an inviolable essence, changing and being changed as they struggle with the modes of oppression that are most unbearably immediate to them. Moreover, insofar as this is ontological, this exchange of essence and manifestation can be understood as a characteristic of all existence. Being is changing clothes in both senses of changing—as an adjective modifying "clothes" and as a verb of which "clothes" is the direct object. The transformative fabric of existence is always changing and exchanging its fabrics for others in an ongoing substitution. In short, "transvestitism comes with difference."[62] Coexistence is a drag show. What remains after God, after nature, after the human, after woman? What remains is just malleable bodies—"plastic bodies"—irreducibly themselves, perhaps taking up the challenge of looking good, multiplying and intensifying differences in performances of exorbitant self-transformation.[63]

A PSYCHEDELIC FEAST

Coexistentialism is queer, feminist, prismatic, and ecological, affirming differences of gender, sexuality, race, and species in the struggles of looking good. Those differences cannot be reduced to biology or social constructions. They are different ways of being, different ways of changing and being changed while simultaneously remaining inviolable. Sexual difference is not privileged, although it is paradigmatic insofar as it exemplifies the overexposure of the intact (inviolable) to contact (violence). The aesthetic challenge of looking good is something that every being takes up, human or nonhuman, even if it does not have eyes to look. Not to be taken too literally, "looking good" is a metaphor for all modes of radiating an appearance while having one's appearance received by others, an exchange of giving, receiving, and transforming. For instance, Morton gives the example of trees, which an artist makes look good while making the artist look good. Myoung Ho Lee has a series of photographs of trees. Each picture of a tree is taken after he puts a large cloth behind it to stage the scene, and the extra clothing makes the sheer specificity of the tree even more apparent. "The fact that you know that it's a stage set, that you can see the wrinkles in the cloth, makes them all the more intense. Like watching someone in drag, you know she or he is performing: queer trees."[64] A queer ecology is a mesh of strange strangers.[65]

As a performance of looking good, sexual difference is not a social construct or a uniquely human endeavor. It names something essential to all beings—inviolable yet vulnerable to others, intimate alterities hiding themselves in plain sight, in and as the clothing of their aesthetic excesses. Vulnerability can turn into violence, when looking good is assimilated into a hegemonic gaze or media spectacle, when difference becomes the site of oppositions and exploitations, when the irreducible specificity of who someone is collapses into how someone appears, when the alterity of the other is lost in the show or, like Kierkegaard's woman, lost in devotion. With its proximity to feminisms of difference, coexistentialism involves commitments to counter such violence in its multiple and intersecting modes, with no universal prescription except to engage the violence you find most pressing in your life, whatever you find most intimately unbearable.

Coexistentialism does not limit itself to countering violence against difference. Affirming sexual differences is about affirming exuberant expression, transvestitism, and the transformation of the senses. Along those lines, sexual difference can be understood as a festive celebration, a feast.

> The link uniting or reuniting masculine and feminine must be horizontal and vertical, terrestrial and heavenly. As Heidegger, among others, has written, it must forge an alliance between the divine and the mortal, such that the sexual encounter would be a festive celebration and not a disguised or polemical form of the master-slave relationship.[66]

The queer, feminist, prismatic, and ecological humans of coexistentialism are thus also festive. Looking good is a party, a feast. This is a reminder that coexistentialism is not primarily a critical endeavor but an affirmative gesture of engagement in the struggles of coexistence, wanting to be in this state of ecological emergency. Furthermore, the festive character of looking good indicates that these queer, feminist, prismatic, ecological hyposubjects are not entirely sober. To some extent, feasts involve intoxication, which has not been entirely welcome throughout the history of philosophy.

The philosophical tradition from Plato to Heidegger displays an ambivalent relationship with drugs and intoxication.[67] There is a discernible tendency for philosophers to privilege sobriety and clear thinking over against the kind of irrational exuberance, difference, and altered states associated with drugs and intoxication. This does not mean that philosophers do not feast, but when they do, it tends to be a very rational feast, feasting first and foremost on speeches (*logoi*). For example, in Plato's *Phaedrus* (227b), Socrates asks Phaedrus to share what he heard when he was entertained by a "feast of speeches."[68] The ensuing discussion addresses the difference between speeches that rely on written aids and speeches performed without writing, and as Derrida has famously elaborated, writing is likened to a drug (*pharmakon*)—an ambiguous substance that is both cure and poison (e.g., writing can cure a deficiency in memory, but it can also facilitate dependency and thus poison one's ability to remember).[69] With ambiguity and dependency, drugs taint the self-sufficient presence and clarity loved by so many philosophers. This relates to the discourse on trance or madness (*mania*) in the *Phaedrus* (265b). Rational *mania* is said to be superior to the other three kinds of *mania* (poetry, divination, and ritual possession), each of which admits of nonrational dependencies (e.g., music and myth) and ambiguities (the vagaries of divination and the morally suspect behavior of ritual possession).

The suppression of *pharmaka* goes hand in hand with the suppression of ecstasies outside the limits of reason.

Drugs contaminate any presence or purity that one might seek at a philosophical feast, and insofar as philosophy in its explicit expression beginning with ancient Greek *philosophia* is defined by its search for such presence or purity, drugs can be said to poison philosophy itself while being suppressed by philosophy. This is related to Malabou's questionable identity as a woman philosopher. As she puts it, "if I'm a philosopher it is at the price of a tremendous violence, the violence that philosophy constantly does to me and the violence I inflict on it in return. My relation to philosophy looks a lot like a constant, fierce quarrel between a man and a woman."[70] This is not a coincidental intersection between drugs and sexual difference. Derrida's analysis connects the suppression of *pharmaka* with the marginalization and exclusion of difference throughout the philosophical tradition. Supplements (e.g., drugs, women, writing, technology) add to and substitute for that which is supposed to be self-sufficient (e.g., health, men, speech, nature), so a desire for purity and self-sufficient presence must systematically suppress its supplements. However, fugitive and recalcitrant, differences return. After God and nature, after the human—man and woman—an alliance of differences is setting the table for a more festive philosophical feast, one that celebrates instead of suppresses differences.

Interrupted by pharmacological supplements, consciousness is never rational but always already altered, identity always already difference, and perception always already hallucination. Instead of reacting against this condition, Nietzsche embraces it. His writings refer positively to uses of drugs, and he sings praises for states of celebration, festivity, and ecstatic inspiration.[71] In a note from "Notebook 34, April–June 1985," Nietzsche describes thinking as *Fest* ("feast"; "festival") and *Rausch* ("intoxication"): "For many people, abstract thinking is fatiguing work—for me, on good days, it is a feast, an intoxication."[72] This is part of Nietzsche's reversal of Plato, recovering the entrancing enthusiasm of the ritual *mania* of Dionysus. To clarify the terminology, it should be noted that "intoxication" is a rather poor translation of *Rausch*, particularly because *Rausch* does not carry a connotation of toxicity, at least not in the direct way that the word "intoxication" actually contains "toxic." *Rausch* is not *Giftigkeit* ("toxicity") or *Trunkenheit* ("drunkenness"). It means "ecstasy" and "transport," and it derives from the Middle High German *rouch* and Old High German *rauh* ("smoke"; "steam").[73] *Rausch* is thus close to *Rauch* ("smoke"), and in that sense it connotes the intemperate mirth of high spirits, floating, drifting, and dissipating into hazes and heavens. The verb *rauschen* refers to the act of making a rushing sound, a whistling, moaning, murmuring. *Rausch* is a rush. Toxicity can be a cause of *Rausch*, but it is neither a necessary nor sufficient condition for *Rausch*.

Moreover, one of the more intoxicating sources of *Rausch* is alcohol, which is not part of Nietzsche's feast.[74]

In Irigaray's statement that an alliance of sexual difference would be a festive celebration, she refers to Heidegger, whose writings occasionally touch upon a "feast" or "festival" of thinking (*das Fest des denkens*), including reference to Nietzsche's account of thinking as feast and *Rausch*.[75] However, Heidegger does not address the place of *Rausch* in the feast, and instead, as Derrida observes, Heidegger defines the feast in terms of "will to power," which Heidegger interprets as a fundamental metaphysical concept from Nietzsche as a metaphysical thinker, missing the exorbitant styles whereby Nietzsche's corpus affirms difference.[76] Keeping Nietzsche within the bounds of metaphysics, Heidegger is also keeping the feast within the bounds of sobriety. This emphasis on sobriety is expressed in one of the poetic dialogues in Heidegger's *Country Path Conversations*, where the festive event of thinking is referred to as a "festival of sobriety."[77] The sober feast of thinking is a way of responding to the event of the open, the gathering of the fourfold—earth, sky, gods, and mortals. Going to the feast is a way of observing one's position in the event—"sober ascertainment [Feststellung]," making a festive statement.[78] The only entrancing enthusiasm is the one that is enthusiastic about sobering up, "enthusing in waiting, whereby we become more waitful and more sober."[79] Preferring sobriety, Heidegger is like someone who says he gets high on life, or rather, high on Being. In addition, with his usual human exceptionalism, Heidegger associates a drinker's lack of sobriety with animals. According to Heidegger, animals do not know how to drink, and that is why their drinking (*saufen*) means "boozing" when applied to humans, i.e., "getting blind drunk."[80] At a feast, who is really more accurately characterized as poor in world, a couple of inebriated humans dancing or a sober philosopher disparaging the dancers for acting like animals?

In sum, Heidegger's feast maintains the proper philosophical suppression of drugs, nonrational ecstasies, nonhumans, in short, difference. He ignores Nietzsche's *Rausch* and never considers what is hidden in Plato's pharmacy.[81] Nonetheless, there is an uncanniness to Heidegger's festival that could perhaps make room for *pharmaka* and other strangers at his seemingly sober feast. The sobriety upon which the interlocutors of his country path conversations are waiting is not the sobriety of an unexamined life. It is weirder than that, more attentive, perhaps even looking good. It is a "strange sobriety."[82] Perhaps it is sobriety without sobriety, an open attentiveness without the habituated, normalized sobriety of the unexamined status quo. Conversely, it could be considered an intoxication without toxicity or a loss of normal consciousness without narcosis. It is this strangeness which opens up Heidegger's festival to difference, opening up his rather restricted sense of the open to a more hospitable relationship with the intimate alterities of coexistence, male and female, human and nonhuman, sobriety and *Rausch*. The strange sobriety seems psychedelic.

Queer, feminist, prismatic, and ecological, hyposubjects are also psychedelic, not only in the aesthetic sense of kaleidoscopically colorful but in the pharmacological sense, referring to a class of drugs, which, moreover, are nontoxic and nonaddictive. Hyposubjects are pharmacological, supplemented, complicit in technologies, mediations, and differences, weakening any pretense of purity or perfect presence. Resisting easy classification in terms of good or bad, cure or poison, and even sobriety or intoxication, psychedelic drugs (e.g., LSD, ayahuasca, and peyote) have radically indeterminate effects, which depend primarily on the mind-set of the person taking the drug and the setting in which it is taken. They have no specific effects apart from catalyzing or amplifying the contents of one's experience, including one's thinking, feeling, sensing, and orientation toward space and time. In a retrospective of decades of psychedelic research since the middle of the twentieth century, Roger Walsh and Charles Grob provide a summary definition of psychedelics.

> The exact nature of psychedelics—what they do, and how they do it—has been a topic for debates and battles almost since their discovery. The names given to these drugs reflect both these battles and the evolution of researchers' understanding. Though originally marketed as "psychotomimetics" and labeled "hallucinogens," their remarkably variable effects soon led Stanislav Grof and others to recognize them as "nonspecific amplifiers" of mental processes that brought previously subliminal experiences to conscious awareness. Hence the name "psychedelic," meaning "mind manifestation." This amplification was particularly true of psycho dynamically charged issues and conflicts, and this was one observation that led to the recognition of their therapeutic potential. More recently, they have been described by the term "entheogen," in recognition of their capacity to elicit spiritual experiences.[83]

As with other practices of looking good, psychedelic experiences involve a multiplication and intensification of differences, exorbitant expressions of different ways of acting and being acted upon. Furthermore, like differences of race, gender, and species, the alterations in consciousness occasioned by psychedelic drugs can be understood in terms of sexual selection. Along those lines, psychedelics have a more prominent place in Darwin's pharmacy than Plato's pharmacy.[84] Psychedelic experiences increase attractiveness, not in the sense that people who take them are necessarily attractive looking, but it can enhance their artistic body transformations, intensifying or differentiating how attractive they sound, like poetic catalysts or rhetorical supplements. Psychedelic experiences enhance capacities for linguistic acrobatics. Avoiding for now any sensational accounts of grandiose trips and heroic journeys, consider a simple experiment with semantic priming.[85] It was shown that doses of psilocybin (the active chemical in "magic" mushrooms) make it possible for people to access semantic associations that are not as readily available in ordinary waking consciousness, i.e., sobriety. When a sober subject is shown a word and asked to read it, the subject can identify the word faster if another related word is shown immediately preceding the word in question. Thus, subjects identify the word "lemon" faster if a related word precedes it (e.g., "sour" or "yellow"). The word "sweet" does not ordinarily facilitate a faster recognition of the word "lemon," because the semantic association between them is too indirect: lemons are sour, the opposite of sweet. Psilocybin increases access to indirect semantic associations, such that the word "sweet" would not impede recognition of "lemon." Rainbow elephants and talking spools of thread start making more sense, and nonsense starts sounding more logical.

A propensity for poetics can be annoyingly verbose, but it can also seem sexy, intensifying emotions and heightening sensations. Of course, much depends on intentions and setting, what you are saying, what your context is, and the taste preferences of your potential mates. In contrast to therapeutic and spiritual potentials of psychedelics, which too often seek to make people well-adjusted or integrated, at the feast of sexual difference psychedelics have a more aesthetic role, intensifying and multiplying linguistic and semantic differences not in order to be healthy, whole, or holy, but to look good, to sound good, to be seductive, alluring, attractive, appealing, experimenting with different ways of remaining exposed to and for intimate alterities. Sociobiological interpretations would argue that sounding good is done for survival, sounding good in order to reproduce. Perhaps it is the other way around, putting reproduction at the service of intergenerational aesthetic transformation. As Grosz observes, a more accurate Darwinian interpretation suggests that such musicality and all artistic expression cannot be understood in terms of survival and natural selection.

> While there is some selection in terms of "fitness," that is in terms of survival, this selection is always negative; it eliminates the unfit or the less fit rather than privileging the fittest. Darwin's argument is that music did not evolve through natural selection but primarily through sexual selection. Music has survived, in other words, not because it is reducible to something useful or practically relevant in everyday life, precisely because it is *not* useful but serves the vaguer purposes of evocative intensification and pleasure. Music develops and survives not because it bestows upon us, its agents and listeners, some direct advantage but because it is pleasing and thus serves to attract others to us and us to others.[86]

As exorbitant intensifications of semantic associations, psychedelics are like sexual selection supplements, which is not to say that it is not also therapeutic and religious. Rather, it is exactly because they are supplements to the specificity of plastic, volatile bodies that psychedelics can function in transformative religious rituals, as adjuncts to psychotherapy, as fun recreational drugs, or as a fetish in the kind of spirituality that is making us stupid, selfish, and unhappy.[87]

As amplifiers of sexual difference, psychedelics can, at least in some cases under specific conditions, enhance one's perception of the differences constitutive of diverse modes of coexistence, catalyzing one's capacity to remain exposed in vulnerable relations with others, creating new spaces of intimate coexistence, looking good. Alongside the queer, feminist, prismatic, ecological hyposubjects of the Anthropocene are the psychedelic subjects. This is not to say that psychedelic drugs have any answers or should be privileged as modes of access to difference. This is to say that the anthropocosmic subjects of the Anthropocene are never neurotypical. They affirm the always already altered character of their brain states, opening onto the insistent question given to all humans, the question of what to do with your brain.[88] What should we do with our plastic, volatile, vibrant bodies?

NOTES

1. Jacques Derrida, *Monolingualism of the Other*, trans. Patrick Mensah (Stanford: Stanford University Press, 1998), 40.
2. Boyer and Morton, "Hyposubjects."
3. Michel Foucault, *The History of Sexuality: Volume 1: An Introduction*, trans. Robert Hurley (New York: Vintage Books, 1990).
4. Elizabeth Grosz, "Naked," in *Encounters with Alphonso Lingis*, ed. Alexander E. Hooke and Wolfgang W. Fuchs (Lanham: Lexington Books, 2003), 130.
5. Ibid., 130–31.
6. Amanda Franklin et al., "The Energetic Cost of Mating in a Promiscuous Cephalopod," *Biology Letters* 8 (2012).
7. Blake, *The Complete Prose and Poetry of William Blake*, 667.
8. Charles Darwin, *The Descent of Man, and Selection in Relation to Sex* (Princeton: Princeton University Press, 1981).
9. Ibid., 2:398.
10. Elizabeth Grosz, *Becoming Undone: Darwinian Reflections on Life, Politics, and Art* (Durham: Duke University Press, 2011), 143.
11. Morton, *The Ecological Thought*, 144n96.
12. Luce Irigaray, *The Way of Love*, trans. Heidi Bostic and Stephen Pluhacek (London: Continuum, 2002), 157.
13. Rose, *Wild Dog Dreaming*, 15, 112.
14. Plumwood, *Feminism and the Mastery of Nature*, 137.
15. Ibid., 167.
16. Søren Kierkegaard, *The Sickness unto Death*, ed. and trans. Howard V. Hong and Edna H. Hong (Princeton: Princeton University Press, 1980), 39.
17. Céline Léon and Sylvia Walsh, eds., *Feminist Interpretations of Søren Kierkegaard* (University Park: Pennsylvania State University Press, 1997). This book is part of a series of anthologies, each of which gives "feminist interpretations of" a specific philosopher.
18. Kelly Oliver and Marilyn Pearsall, eds., *Feminist Interpretations of Friedrich Nietzsche* (University Park: Pennsylvania State University Press, 1998). Nancy J. Holland and Patricia Huntington, eds., *Feminist Interpretations of Martin Heidegger* (University Park: Pennsylvania State University Press, 2001).

19. Simone de Beauvoir, *The Second Sex*, trans. Constance Borde and Sheila Malovany-Chevallier (New York: Random House, 2009), 283.
20. Ibid.
21. Ibid., 7.
22. Catherine Keller, *From a Broken Web: Separation, Sexism, and Self* (Boston: Beacon Press, 1986), 20.
23. Ibid.
24. I focus on Irigaray here and not Pluwmood because the former is more thoroughly and explicitly engaged with the legacies of existentialism.
25. Luce Irigaray, *I Love to You: Sketch for a Felicity within History*, trans. Alison Martin (New York: Routledge, 1996), 107.
26. On Nietzsche's question of "*how one becomes what one is*," see Nietzsche, *Ecce Homo*, 34. Cf. Nietzsche, *The Gay Science*, 219 (#270).
27. Irigaray, *This Sex Which Is Not One*, 214.
28. Ibid., 79.
29. Catherine Malabou, *Changing Difference: The Feminine and the Question of Philosophy*, trans. Carolyn Shread (Cambridge: Polity Press, 2011), 10.
30. Luce Irigaray, *An Ethics of Sexual Difference*, trans. Carolyn Burke and Gillian C. Gill (London: Continuum, 2004), 6.
31. Luce Irigaray, *Luce Irigaray: Key Writings*, ed. Luce Irigaray (London: Continuum, 2004), 178.
32. Irigaray, *I Love to You*, 35.
33. Luce Irigaray, "The Question of the Other," trans. Noah Guynn, *Yale French Studies* 87 (1995): 19.
34. Grosz, *Becoming Undone*, 100.
35. Ibid., 107.
36. Luce Irigaray, "Women's Exile," trans. Couze Venn, *Ideology and Consciousness* 1 (1977): 69.
37. Grosz, *Becoming Undone*, 109.
38. Ibid., 138. As Darwin says, "the differences between the races of man, as in colour, hairyness, form of features, &c., are of the nature which it might have been expected would have been acted on by sexual selection." Darwin, *The Descent of Man*, 2:250.
39. Grosz, *Becoming Undone*, 139.
40. Morton, "X-Ray," 311.
41. Jeffrey Jerome Cohen, ed., *Prismatic Ecology: Ecotheory Beyond Green* (Minneapolis: University of Minnesota Press, 2013).
42. Timothy Morton, "They Are Here," in *The Nonhuman Turn*, ed. Richard Grusin (Minneapolis: University of Minnesota Press, 2015), 173.
43. Grosz, *Becoming Undone*, 109–10.
44. Some of the prominent figures in this trajectory of feminist theory include Grosz, Haraway, Claire Colebrook, Karen Barad, and Stacy Alaimo, each of whom contributes a chapter to the following anthology: Stacy Alaimo and Susan Hekman, eds., *Material Feminisms* (Bloomington: Indiana University Press, 2008). On Grosz's "corporeal feminism," see Elizabeth Grosz, *Volatile Bodies: Toward a Corporeal Feminism* (Bloomington: Indiana University Press, 1994). Material and corporeal feminism overlap with non-anthropocentric orientation of ecofeminist perspectives. Mary Phillips and Nick Rumens, eds., *Contemporary Perspectives on Ecofeminism* (New York: Routledge, 2015).
45. Catherine Malabou, *Plasticity at the Dusk of Writing: Dialectic, Destruction, Deconstruction*, trans. Carolyn Shread (New York: Columbia University Press, 2010), 61.
46. Ibid., 67.
47. Malabou, *Changing Difference*, 138.
48. Ibid., 34.
49. Ibid., 28.
50. Ibid., 96.

51. Ibid.
52. Ibid., 93. Malabou is drawing on Derrida's sense of "remains," as in what remains after the conclusively comprehensive philosophy of Hegel, after absolute knowledge. See Jacques Derrida, *Glas*, trans. John P. Leavey Jr. and Richard Rand (Lincoln: University of Nebraska Press, 1986).
53. Malabou, *Changing Difference*, 93.
54. Ibid., 94.
55. Ibid., 140.
56. Ibid., 98.
57. On the dust that remains—just dust—see chapter 3 ("After God, After Nature").
58. Malabou, *Changing Difference*, 99.
59. Timothy Morton, "Treating Objects Like Women: Feminist Ontology and the Question of Essence," in *International Perspectives in Feminist Ecocriticism*, ed. Greta Gaard, Simon C. Estok, and Serpil Oppermann (London: Routledge, 2013), 56.
60. I discuss the specific differences between new materialist and object-oriented perspectives in chapter 9 ("Becoming Worldly").
61. Ibid., 36.
62. Ibid., 37.
63. I borrow the phrase "plastic bodies" from Tom Sparrow's reconfiguration of phenomenological conceptions of embodiment through an engagement with plasticity. Tom Sparrow, *Plastic Bodies: Rebuilding Sensation after Phenomenology* (London: Open Humanities Press, 2015).
64. Morton, *Realist Magic*, 144.
65. Morton, "Guest Column: Queer Ecology," 275–77.
66. Irigaray, *An Ethics of Sexual Difference*, 17.
67. For a lyrical, perhaps intoxicated, account of the ambivalent relationship between philosophy and intoxication, see Jean-Luc Nancy, *Intoxication*, trans. Philip Armstrong (New York: Fordham University Press, 2016).
68. Another feast of speeches is referenced in Plato's *Timaeus* (20c).
69. Jacques Derrida, *Dissemination*, trans. Barbara Johnson (Chicago: University of Chicago Press, 1981), 63–171.
70. Malabou, *Changing Difference*, 140–41.
71. "If one wants to get free from an unendurable pressure one needs hashish." Nietzsche, *Ecce Homo*, 31.
72. Friedrich Nietzsche, *Writing from the Late Notebooks*, ed. Rüdiger Bittner (Cambridge: Cambridge University Press, 2003), 9.
73. David Goldblatt, *Art and Ventriloquism: Essays* (New York: Routledge, 2006), 183n21.
74. "I cannot advise all *more spiritual* natures too seriously to abstain from alcohol absolutely. *Water* suffices." Nietzsche, *Ecce Homo*, 23.
75. Martin Heidegger, *Nietzsche: Volumes One and Two*, trans. David Farrell Krell (San Francisco: Harper SanFrancisco, 1991), 5. Cf. Heidegger, "The Origin of the Work of Art," 2.
76. Jacques Derrida, "Interpreting Signatures (Nietzsche/Heidegger): Two Questions," in *Dialogue and Deconstruction: The Gadamer-Derrida Encounter*, ed. Diana P. Michelfelder and Richard E. Palmer (Albany: SUNY Press, 1989), 68–69.
77. Martin Heidegger, *Country Path Conversations*, trans. Brett Davis (Bloomington: Indiana University Press, 2010), 89.
78. Ibid.
79. Ibid., 94.
80. Ibid., 88.
81. Plato didn't just write about drugs in his literary corpus. His body—his literal corpus—was on drugs, too. One can safely assume that Plato consumed some wine during his lifetime. Of course, alcohol is not what people typically have in mind when they wonder if Plato was on drugs. They are wondering about whether there was something more unusual, intense, or psychoactive. The wine of Plato's Athens was

indeed more psychoactive, quite unlike today's alcoholic beverages, which are mainly alcohol and maybe some added juices, flavoring, carbonated water, etc., but no added psychoactive plants (absinthe is a notable exception). For a detailed source of information on the properties of the kind of wine imbibed in Plato's Athens, see Michael Rinella, *Pharmakon: Plato, Drug Culture, and Identity in Ancient Athens* (Lanham: Lexington Books, 2011). There is another, much more speculative, hypothesis that, along with psychoactive wine, Plato might have taken a psychedelic drink in an initiation into the esoteric religious cult of the Eleusinian mysteries. The argument claims that the ritual drink (*kykeon*) ingested during that initiation was made using barley that was infected with a fungus (ergot), which contains an alkaloid analogous to LSD. Although not entirely unreasonable, that hypothesis seems quite unlikely, since it would be difficult to intentionally infect the barley with exactly the right amount of fungus: too little and the initiates would not have the initiatory experience, too much and they would be poisoned, although perhaps that risk was part of the initiation. Regarding this speculative hypothesis, see Robert Gordon Wasson, Albert Hofmann, and Carl A. P. Ruck, *The Road to Eleusis: Unveiling the Secret of the Mysteries*, 30th ann. ed. (Berkeley: North Atlantic Books, 2008).

82. Heidegger, *Country Path Conversations*, 94.

83. Roger Walsh and Charles S. Grob, "Conclusion: What Did These Elders Learn, and What Can We Learn From Them?," in *Higher Wisdom: Eminent Elders Explore the Continuing Impact of Pyshcedelics*, ed. Roger Walsh and Charles S. Grob (Albany: SUNY Press, 2005), 242.

84. In *Darwin's Pharmacy*, Richard Doyle outlines the evolutionary interconnectedness of humans and psychedelics in the process of sexual selection. Emphasizing that interconnectedness, Doyle coined the term "ecodelic" to describe this class of drugs. Richard. M. Doyle, *Darwin's Pharmacy: Sex, Plants, and the Evolution of the Noösphere* (Seattle: University of Washington Press, 2011), 31.

85. Manfred Spitzner et al., "Increased Activation of Indirect Semantic Associations under Psilocybin," *Biological Psychiatry* 39.12 (1996).

86. Elizabeth Grosz, *Chaos, Territory, Art: Deleuze and the Framing of the Earth* (New York: Columbia University Press, 2008), 35–36.

87. Regarding the argument that spirituality is making us stupid, selfish, and unhappy, see chapter 3 ("After God, After Nature"). See also David Webster, *Dispirited.*

88. These are among the questions raised by Malabou in her account of neuroplasticity and its ethico-political implications. Catherine Malabou, *What Should We Do with Our Brain?*, trans. Sebastian Rand (New York: Fordham University Press, 2008).

NINE

Becoming Worldly

Ecological emergency shows up everywhere, experienced locally, covered in the media, debated in politics, insinuated and implicated in everyday conversations. It is evident in superstorms, drinking water contamination, record temperatures, oceanic gyres of plastic, floods and droughts, virus outbreaks, air pollution, food shortages, migrations (human and nonhuman), habitat destruction, deforestation, desertification, and so on, in an apparently countless litany of ecological problems. One is never done counting the actually existing entities involved in these emergency conditions. Who counts? Who does the counting, and who is counted, and how much and for what does each count? Every being woven into the mesh of planetary coexistence resonates with the unheard of, issuing the compelling call of that which is wholly other. In an ecology of alterities, who counts? Who are we? This is Derrida's opening questioning in *The Politics of Friendship* by repeating the impossible question, "How many of us are there?"[1] Others go and others come, and if "every other is altogether other" (*tout autre est tout autre*), then there is always more otherness to come as others come and go.[2]

What makes the current ecological emergency an emergency is not simply that there are countless others undergoing uncertain and dangerous transformations as humans become increasingly interwoven with the life, land, air, and water of Earth. Unprecedented uncertainty and danger are part of the emergency conditions of the current situation, but those conditions would not be so pressing, critical, and urgent if they were simply "those" conditions, those problems "over there," "away," in the "environment." The emergency is an emergency precisely insofar as it is forcing nonhumans into human lives, ruining the neatly divided categories that kept nature distinct from culture, science distinct from politics, facts distinct from values, subjects distinct from objects, selves distinct from others. The emergency that is urgently emerging is the swarm of coexistence, the crowd of intimate alterities, in which the human loses its exceptionally hypertrophic subjectivity and remains exposed, vulnerable, different, intimately intertwined with countless others. This is not some metaphysical theory of intimacy. It is coexistence. "The current ecological emergency," as Morton correctly observes, "confronts us with the non-metaphysical intimacy of nonhuman beings," which is to say, the emergency "just is the existence and coexistence of things."[3] Accordingly, responses to ecological emergency are not about solving problems and facilitating sustainability transitions, or at least not primarily so. To respond to the emergency is to encounter intimate alterities, which means letting go of fantasies about something underneath us or around us called "nature," "environment," or "world," which needs to be saved by "us"—some select group of hypersubjects, i.e., way-too-able-bodied humans.

There is no group of humans that need to save nature over there and around us. This is not because God will do the saving instead of humans, and it is not because we humans are nature, such that we are actually nature saving ourselves. It is because those categories do not make sense anymore. There is no world to save, not a world separate from humans or one that is identical with humans. As Nancy says, there is "no longer any sense of the world," not in the sense of "a *mundus*, a *cosmos*, a composed and complete order (from) within which one might find a place, a dwelling," and "no longer the 'down here' of a world one could pass through toward a beyond or outside of this world."[4] After God, after nature, after humans, coexistentialism is also after the world. In a situation that would surely be horrifying to Heidegger, human existence is more like being-in-the-dust than being-in-the-world. The worldlessness (*Weltlosigkeit*) that he saw in stones now characterizes human existence. "Contra Heidegger, we coexist in lacking-a-world."[5] The point is not that humans have lost their ecstatic openness but that such openness is not attributable to some horizon called "world." Once that horizon disappears, it is much easier to recognize that openness is not an exclusively human trait. Even stones are open, remaining exposed to contact with others.

In the ruins of the world, all that remains is just the dust of this planet, the shimmering singularities of the swirling swarm. In this dust, humans are learning to become human again and for the first time. Learning to become human—becoming anthropocosmic in the Anthropocene—means learning to recalibrate the activity of the human subject so that it is less hyper and more sincerely different—strange, feminist, queer, prismatic, psychedelic, weak, lame, hypocritical, ecological. Remaining exposed, open and vulnerable to intimate alterities, humans no longer inhabit the spheres and globes of classical and modern worlds, and in the climate crisis that remains in the ruins of those worlds, a planetary immune system remains to be constructed. As Nancy points out, construction today is not a matter of finding a new foundation. Nor is it a matter of reconstruction after deconstruction ("contrary to the incessantly repeated petition addressed to 'deconstructionists': will you reconstruct already?"); rather, "*struction*" as such is the issue.[6] Struction, from *stuere* ("to pile up," "to heap") is about heaping, piling, amassing, assembling, stretching out, extending.

To think about coexistence, one must think about struction. It means thinking about the heap of beings, *partes extra partes*, without a background, organization, foundation, or structure. The heap is the site of one of the paradoxes of coexistence, the sorites paradox, which indicates that a piling up of beings with other beings does and does not constitute a pile. Morton summarizes the paradox.

> I have a heap of sand. I remove one grain. It's still a heap. I can keep going until I have just one grain left. It's not clear at what point, if at all, it stops being a heap. Try it in reverse. If I have one grain of sand, it doesn't make a heap. If I add another grain, it doesn't make a heap either. Now I can carry on with the same reasoning process indefinitely—so I never get a heap, no matter how many tens of thousands of grains I pile up.[7]

The status of this paradoxical heap without heap opens up a rift between ecological approaches to existentialism, namely the ecological existentialism of Rose and the coexistentialism of Morton. Both approaches share a common orientation after the end of the hegemonic homogeneity of "world," and contra Heidegger, Rose and Morton celebrate the loss of that lonely, anthropocentric world. It was, as Irigaray observes, a world of anesthetizing fantasies that block out the sexual and elemental specificities of bodies, preventing the hurt from hurting and thus preventing the emergency from emerging.

> You have built an anaesthetic world. But when our greatest pains or greatest joys are abolished by calculation, is that not the worst destruction? . . . And there is no remedy. Your world of anaesthetics kills insensibly. Irremediably. The more you go on producing fantasies, your preferred anaesthetics, the greater the danger: avoiding the passage through suffering which could still save.[8]

The variety of thinkers affiliated with coexistentialism or ecological existentialism celebrate the loss of that anesthetic world, which had no room for feminist, queer, prismatic, psychedelic, ecological, and elemental differences. Nonetheless, there is a world of difference between them. In the ecological existentialism of Rose, this heaping can be thought of as worlding, an ethical mode of relating to others in continuity and difference, and for the coexistentialism of Morton, this heaping is thought of in terms of objects who relate without world, worlding, environment, or any such ambience that provides a middle ground between objects.

Tracing this rift, I discuss Morton's account of the end of the world and outline an object-oriented account of ecology without world. By returning to Derrida and Nancy, I consider the possibility that, while coexistence is indeed without world, and ecology is after the end of the world, the world remains and continues returning in the coming and going of beings. It is through the coexistential struggle to bear the unbearable weight of the world that practices for the construction of a planetary immune system can be sought out (see chapter 10, "*Askesis*"). It is not a struggle for a world that could contain, ground, or mediate planetary coexistence. It is a struggle for a world that is coexistence.

ECOLOGY OF THINGS, ECOLOGY OF NOTHINGNESS

For Morton, the end of the world already happened. Indeed, it has a relatively particular date.

> It was April 1784, when James Watt patented the steam engine, an act that commenced the depositing of carbon in Earth's crust—namely, the inception of humanity as a geophysical force on a planetary scale. Since for something to happen it often needs to happen twice, the world also ended in 1945, in Trinity, New Mexico, where the Manhattan Project tested the Gadget, the first of the atom bombs, and later that year when two nuclear bombs were dropped on Hiroshima and Nagasaki.[9]

The end of the world is the birth of the Anthropocene, a protracted event that emerged between the beginning of the industrial age and the beginning of the atomic age. What emerged in this event is an emergency. "When the charm of *world* is dispelled, we find ourselves in the emergency room of ecological coexistence."[10]

This period of ecological emergency corresponds to the same period of time that marks the death of God and death of nature. Indeed, the deaths of God and nature are precisely losses of world, as God and nature functioned as the grounding, organizing, orienting functions of the world, typically in respectively transcendent and immanent spheres. The end of the world is the ruin of the metaphysical and physical climate of planetary coexistence. Nietzsche understood this coinciding loss of metaphysical truth (God) and physical appearance (nature) in the loss of the world. "*Along with the true world, we have also done away with the apparent!*"[11] The multicultural idea of many worlds sharing one Earth simply pluralizes the true world, grounding the diversity of worlds in an immanent and objective Earth, thus recapitulating nature-culture dualisms and failing to come to terms with this loss of the apparent *and* the true, this loss of Earth and world in the dust of this planet.[12]

Along with the deaths of God and nature, the loss of the world also corresponds with the end of humanism, as the *anthropoi* of the Anthropocene are rendered weak, lame, and hypocritical by the hyperobjects coming forth in this era, massively distributed entities that defy human understanding, such as global capitalism, black holes, and climate change. Hans Jonas observed something like this in the late 1970s, reflecting on discoveries in "the science of ecology," which indicate "that the nature of human action has *de facto* changed, and that an object of an entirely new order—no less than the whole biosphere of the planet—has been added to what we must be responsible for," an object that is "dwarfing all previous objects of active man!"[13] The term Anthropocene is capitalized because it names a geological epoch or at least a boundary event. It does not imply a moral anthropocentrism or androcentrism. It implies that any occupation of a hierarchical center is swarmed with hyperobjects also vying for center stage. There is no periphery against which humans can claim central status. That periphery is a swarm of beings, plants and animals, bacteria and fungi, mountains and rivers, refrigerators and computers, and other humans excluded from central status due to sex, gender, race, age, ability, religion, geography, etc. It is this horizon that has been lost with the end of the world and the advance of hyperobjects.

The end of the world is more than the end of the spheres of religious traditions and classical philosophies, which is to say, more than the end of premodern superstitions and speculations. Furthermore, the end of the world means more than recognizing that metaphysical cosmologies and scientific determinations of the universe are merely abstractions, and that the world they describe thus does not actually exist in distinction with the world opened up in concrete experience. Not only have the worlds of philosophy, religion, and science ended, but so too has the world that opens up in experience. That world is described well by Husserl.

> The word is pregiven to us, the waking, always somehow practically interested subject, not occasionally but always and necessarily as the universal field of all actual and possible praxis, as horizon. . . . The world, on the other hand, does not exist as an entity, an as object, but exists with such uniqueness that the plural makes no sense when applied to it. Every plural, and every singular drawn from it, presupposes the world-horizon.[14]

If it ever existed, that horizon is gone. If you are looking for a horizon you will only find things, entities, beings, piling up, heaping, opening up spaces and times for interactions with other objects. There is no horizon or mediating field. Objects do their own mediating, translating, and interpreting. They do not need to outsource that work to a horizon, ground, or medium at large. Husserl is right that the world does not exist as an entity. It is not a mega-entity containing other entities, nor is it a totality of entities, as if entities could be consistently counted and totalized. The world does not exist as an entity, not because it exists as a horizon, but because it does not exist at all. "The sensation of *world*," for Morton, "is the false consciousness of gaps and backgrounds between and behind things."[15]

What does exist? For Levinas, existence *sans monde* ("without a world") is an elemental anonymity, which still functions like a background.[16] However, that is thus not sufficiently *sans monde*. Without world, without background, ground, foundation, nature, without nature, there is just dust, a swarm of singularities. In short, there are things. After the end of the world, ontology no longer has a world, nature, omnipotent God, hypersubjective human, or any such mega-being that functions as a ground, source, foundation, or sole locus of meaning and value in existence. What remains is an ontology oriented toward swarming singularities. Morton calls this ontology of things "object-oriented ontology" (OOO), using the term "object" interchangeably with "being" or "entity," not as the opposite of a subject.[17] Facilitating the birth of object-oriented ontology out of the spirit of the Anthropocene, Morton follows Graham Harman, who began using the phrase "object-oriented philosophy" at least as early as 1999.[18] The term "object-oriented ontology" was coined by Levi Bryant in 2009, and OOO began emerging as a movement the following year with a series of conference and publications, which featured Harman and Bryant with the other founding figure of OOO, Ian Bogost. Bogost and Harman draw heavily on phenomenology to develop their respective versions of OOO.

To be clear on the frequently misleading terminology, the objects of OOO are not opposed to subjects. The status of these terms has changed throughout the history of philosophy, and the use of "object" in this context should not be confused with its modern and postmodern definition as something passive, inert, lacking agency or subjectivity, lacking meaning or value. Object-oriented approaches hold up this discarded and dejected category of existence in an effort to do justice to all things in their exorbitant specificity, understanding and affirming things as intimate alterities with different capacities for acting and being acted upon. Throughout the history of philosophy, objects often have been overlooked and underlooked. In Harman's terms, views of reality frequently fall into the tendency of "undermining" things (reducing them to constituent parts or underlying processes and networks), or "overmining" them (reducing them to social constructs or to appearances or correlations to humans).[19] It is easy to find examples of each.

Social constructionism and idealism overmine things by reducing them to questions of human language, experience, and culture. In that view, things are not real, they are just effects of anthropomorphic projections, language games, and sociopolitical power. A glass of water is only called a glass of water based on the way human cognition weaves together sense impressions of hard, smooth, translucent, etc., hypothesizing the existence of a stable entity based on one's habitual sense of those impressions, or you call it a glass because of your tacit acceptance of grammatical structures within which a particular nominalized referent (glass) can be designated and described in a genitival or possessive relationship to its contents (water). The glass could be anything whatsoever, depending on what you and your society make of it. Aside from human interpretation, it is as good as nothing, or whatever.

A different reduction of things can be found in materialism and process-relational philosophy, which undermine things by reducing them to underlying activities and relational networks. In that case, a glass of water is an abstract way to refer to what is actually just a composite of atoms (Democritus) or actual occasions (Whitehead), or in the language of newer versions of materialism and process-relational philosophy, a glass is a whirring assemblage of physicochemical occasions generating apparent continuity over time through complex and stochastic recursive interactions between its parts internally and its external conditions. To further emphasize that things are reducible to the relations they are *within*, Karen Barad coined the word "intra-action" to replace interaction, the latter of which connotes relations *between* (inter-) things and thus does not sufficiently reduce things to their constitutive relationships.[20] This is not unlike Haraway's category of companion species, which follows Barad's concept of intra-action when "insisting on the relation as the smallest unit of being and of analysis."[21] On one hand, as constitutive intra-action, such a theory of relation undermines things. On the other hand, if the relation is the couple, such that $1 + n$ objects are the smallest unit of being and of analysis, then Haraway and object-oriented ontology would not be in disagreement.

To account for things in their irreducible specificity, Harman follows philosophers like Aristotle and Leibniz, for whom substances are the primary reality, but Harman erases the dotted lines that they put around substances.[22] For Harman, things do not have to be natural, but can also be artificial. An artificial thing is not reducible to its natural predecessor, e.g., a table cannot be reduced to a tree. Likewise, things can be simple and composite. A string and a person are things, and so are violins and orchestras. Furthermore, every single thing harbors a radical alterity, a withdrawal that Heidegger articulates in his famous tool analysis in *Being and Time*. A consideration of your experience of an everyday thing is sufficient to disclose the irreducible singularity whereby a thing remains ungraspable and inviolably other even when I touch it and use it. When I use a hammer, the reality of the hammer is invisible yet exposed to relationships. This is its readiness-to-hand (*Zuhandenheit*), which is not reducible to presence-at-hand (*Vorhandenheit*), which is the hammer that becomes present in my relations—the hammer I notice, such as when it is broken or somehow becomes an object of my explicit concern, whether an object of theory or discourse. In short, the real hammer is right here, in my hand, yet never reducible to presence-at-hand, never accessible within the horizons of theory or practice. It is withdrawn, and in the interval of that withdrawal, it manifests its definitive action, its readiness, which means that I can use it and relate to it precisely insofar as it is not present for any theory, any practice, in short, any relation.

> The ready-to-hand is not grasped theoretically at all, nor is it itself the sort of thing that circumspection takes proximally as a circumspective theme. The peculiarity of what is proximally ready-to-hand is that, in its readiness-to-hand, it must, as it were withdraw [*zurückzuziehen*] in order to be ready-to-hand quite authentically. That with which our everyday dealings proximally dwell is not the tools themselves [*die Werkzeuge*].[23]

To put it another way, the real thing itself is not present, yet in its withdrawal it exhibits the pull of an attraction (*Bezug*) whereby one can relate to the thing indirectly, aesthetically. The thing never becomes present, but we relate to it indirectly insofar as its hiddenness is alluring, pulling us into relation. Moreover, the withdrawn feature of the tools themselves extends to all things, not just to what is conventionally included in the category of tools. A hammer, a hydroelectric dam, a jug, a bee, a theory, an ecosystem, an army, a human, and a Greek temple all share the same tension between withdrawal and presence.

With his concept of withdrawal, Heidegger pulls phenomenology into deeper attention to things themselves as they exist outside of the horizons of human access. However, Heidegger's realism is not realist enough. It still bears anthropocentric tendencies, as the withdrawn reality of things tugs only on human concern, with nonhumans left with no meaningful contact with other beings. Harman integrates Heidegger's insight into withdrawal while overcoming Heidegger's anthropocentrism by integrating insights from Whitehead and Latour, for whom human-world relations are not central but are merely a special case of relations between any actual entities (Whitehead) or actors (Latour). What occurs between humans is different only in degree, not in kind, from what occurs between any objects. Bringing those insights together, Harman posits a plurality of actors, each with an inaccessible core—a heart of alterity that exceeds all relations. This tension between inaccessible reality (withdrawal) and accessible relations (presence) occurs in all object-to-object interactions (including an object's interactions with itself).

Beings are integral units withdrawing not just from human contact but from all contact. Moreover, substances are not opposed to relations or processes. Rather, every object is a substance insofar as it is a unit with an unknown (withdrawn) reality (a real object, a strange stranger), and every object is processual and relational (a sensual object, the mesh) insofar as it is composed of interactions and relations with its parts and with other objects.[24] Everything is a strange stranger in the mesh of coexistence. Substance is not a primordial ontotheological ground. For Harman, there is no reason to preclude an infinite regress of objects within objects within objects, and even if there is some tiniest object or substratum, it cannot explain the withdrawn reality of any of the objects of which it is a part or constituent layer. Consider an example: fire burning cotton. The fire enters into a relationship with the cotton by translating its alluring qualities of flammability into its own horizon of combustion. The fire does not directly touch the cotton itself. They touch indirectly, through their translations of one another, their aesthetic encounters of acting and being acted on. Through this indirect touch, the cotton is consumed in flames without the flames ever touching the withdrawn core of the cotton. Just as human contact (theoretical and practical) never exhausts the profound being of real fire and cotton, "the fire and cotton also fail to make full contact with *each other* when they touch, despite their uniting in a bond of destruction."[25] Also, the cotton does not make direct contact with its own parts (the fibers, molecules, atoms, etc.). In short, all causality happens vicariously, in such a way that things touch each other without touching at all, enmeshing while remaining irreducibly other. Things interact through their aesthetic encounters—translations, interpretations, looking good—thus making contact while leaving one another intact.

Harman's vicarious cause or "indirect cause" is a secularized version of the causal theory of occasionalism, which is a philosophical position that first emerged in Ash'arite schools of Islamic theology, according to which objects never directly relate to one another causally but are mediated by God.[26] To say that fire burns cotton denies God's omnipotence, delegating God's power to other causes, i.e., natural causes, called "secondary causality" in scholastic terminology.[27] The fire burns the cotton indirectly through the occasion of God's intervention. Of course, in an animism without animism, after God and nature and world, it is untenable to posit a mega-being as the sole guarantor of all causal interactions. For object-oriented ontology, the vicarious cause comes not from an ontotheological superpower but from the aesthetic exuberance of things themselves, the sensual appearances whereby strange strangers act on and are acted on by others. The fire acts on the reality of the cotton indirectly through a pyromorphic translation of the cotton's appearance, destroying the cotton without ever grasping its irreducible strangeness. As Morton observes, this nontheistic or non-ontotheological version of occasionalism can be thought of in terms of Mahayana Buddhist theories of causality, such as that articulated by Nagarjuna.[28] In that case, it is not God but the emptiness (*śūnyatā*) of things that allows beings to touch without touching.

Emptiness is not nothing. As the Kyoto school philosopher Keiji Nishitani puts it, *śūnyatā* is not "nihility," which "renders meaningless the meaning of life."[29] Emptiness is reality, actual existence.[30] It is neither life nor its negation in death, as it is the withdrawn reality of living and dying. "In this sense, such a mode of being might be termed life-*sive*-death, death-*sive*-life."[31] Perhaps "undead" would work for short. It is neither being nor its negation, but is "absolutely being-*sive*-nothingness, nothingness-*sive*-being."[32] In the famous formulation of the Mahayana scripture, *Prajñāpāramitāhṛdaya* (often referred to as *The Heart Sutra*), "form is emptiness, and emptiness is form."[33] Not nothing, emptiness is the reality of what appears. It is exactly the thinghood of something, whereby a thing is irremediably itself, overwhelmingly singular, displaying itself and thus entering into contact (display) while remaining intangible.

With no background, foundation, or world behind the scenes, things are just things, each so exorbitantly itself that you remain empty handed even while grasping it. Morton relates this to the Christian theologian Paul Tillich, who articulates a distinction between *oukontic* ("not-being") and *meontic* ("un-being") senses of nothing, the former as a nihilistic void, a lack of existence that Morton calls "not even nothing," and the latter as the palpable, horrifying, and fascinating strangeness of things.[34] Emptiness is *meontic* nothingness.

> *Nothingness,* rather than absolutely nothing: and this nothingness pervades things like myriad cracks in the shell of a boiled egg. Because a thing withdraws, it disturbs us with an excess over what we can know or say about it, or what anything can know or say about it—this excess is a nothingness, not absolutely nothing, but not something to which one can point.[35]

Not nothing, the emptiness or nothingness of things is intangible not because it is a lack or deficiency but because its exorbitant display is so in your face that you can never figure out if you are touching the thing itself or the thing as it appears to you. This exorbitant display means that "nothingness is strangely *physical*."[36] When a raindrop hits my head, its emptiness is what makes it possible for it to get me wet, touching me while remaining intact, so thoroughly itself that even when it is touching me it still remains entirely other, unassimilable into any aspect of my experience or any aspect of any other being.

The nothingness of an object is an intact alterity, such that encountering objects is a matter of encountering strange strangers, alien substances. Object-oriented philosophy can thus be described with Ian Bogost's term, "alien phenomenology."[37] For Bogost's alien phenomenology, the alien realities under investigation are all things, human, nonhuman, or otherwise. The orientation toward the alterity of all objects calls for a weirder and more realist phenomenology, which uses speculative metaphysics and metaphor to account for things exceeding human access. A unique difference in Bogost's object-oriented method from Harman's or Morton's is that Bogost is craftier, which is to say, Bogost (a video game designer) focuses more on what he calls "carpentry," which includes operations and procedures for approaching and crafting particular objects.[38] Alien phenomenology is a crafty version of OOO, making it particularly relevant for addressing ecological issues that call for practical procedures and for the creation of new policies, designs, and institutions.

With the object-oriented approaches like those of Morton, Harman, and Bogost, ecology is much more than the study of relationships between organisms and their environments. It becomes much wider and weirder than that, orienting itself toward the alien alterity of all things, an ecology of nothingness. This is a thoroughly non-anthropocentric approach to ecology insofar as it affirms the agency and irreducible alterity of nonhumans, accounting for different agencies and alterities not only in human-to-human or human-to-nonhuman relationships but also in nonhuman-to-nonhuman relationships. Furthermore, an object-oriented approach to ecology is ethical as well as ontological. The alterity of an object pulls on other objects. Things attract and compel one another. Just like Levinasian phenomenology finds ethical imperatives issuing from the alterity of the face of the human other, alien phenomenology finds ethical imperatives issuing from the alterity of every object, and not simply *from objects to humans,* but *from objects to other objects* of all kinds, such that there is something like a proto-ethical relationship between elephants, between atoms, between a hammer and a nail, between fog and the Golden Gate Bridge, between the sun and Jupiter. The sur*face* of the floor expresses an imperative that compels me to responsibility while also expressing imperatives to chairs sitting on it, the cat sleeping on it, the lights that are shining on it, and the linoleum and planks of timber and metal alloy that compose its parts.

Levinasian alterity always refers back to a human subject, but when the other and the subject are both nonhuman, the human becomes entangled in an immense mesh of ethical relationships, such that ethics itself begins to look like a hyperobject. Bogost considers the massive challenges of an object-oriented approach to ethics.

> We can imagine scores of bizarro Levinases, little philosopher machines sent into the sensual interactions of objects like planetary rovers. Their mission: to characterize the internal, withdrawn subjectivities of various objects, by speculating on how object–object caricatures reflect possible codes of value and response. Object ethics, it would seem, can only ever be theorized once-removed, phenomenally, the parallel universes of private objects cradled silently in their cocoons, even while their surfaces seem to explode, devour, caress, or murder one another. . . . Ethics itself is revealed to be a hyperobject: a massive, tangled chain of objects lampooning one another through weird relation, mistaking their own essences for that of the alien objects they encounter, exploding the very idea of ethics to infinity.[39]

If environmental ethics is an extension of classical and modern ethics, object-oriented ecology makes ethics far stranger. Ethics explodes into intimate immensity as we find ourselves amid a vast and tangled chain of compelling objects. Ethical qualities are withdrawn, inaccessible, empty, making it impossible to decide with any certainty how to respond to others, how to count who counts. You cannot jump over your own shadow. Ethics is contained in the limits of a certain "self-centered practice."[40] That spells an inevitable anthropocentrism for object-oriented ecology, albeit an anthropocentrism where humans are no longer sovereign among beings, that is, an anthropocentrism after the deconstruction of center/periphery binaries in the Anthropocene. Anthropocentric hypersubjects gives way to the anthropocosmic hyposubjects.

In the uncanny context of ecological emergency, the self-centered human can no longer hold onto the fantasy of being an exclusive center. As the center and periphery are rendered untenable, so is the world and all the "middle objects" (e.g., nature, place, space, environment, horizon) that are supposed to provide the glue that holds everything together, glue for pasting over the uncontrollably exuberant nothingness of things.[41] In unbearably intimate proximity to hyperobjects, humans are undergoing the uncanny realization that we are all strange strangers among countless others, coexisting after the end of the world.

CARRYING ON: THE STRUGGLE

Embracing the uncertainty and connectivity of life in the Anthropocene, Rose's ecological existentialism draws on a Levinasian sense of alterity, which she redistributes to include all of the human and nonhuman constituents of planetary coexistence. "Ecological existentialism enjoins us to live within the dynamics, and to pour our love into this unstable and uncertain Earth."[42] The relationships between others are relations of "becoming with," where "becoming" is not an underlying process (which would undermine actually existing others) but refers to the contingency and ongoingness of existence. Becoming with one another, terrestrial others produce shared meanings and spheres of intimacy spheres, which Rose calls "world making": "In becoming with others we bring forth worlds of action and meaning, with varied possibilities for life and for death," which means that all beings are "engaged in world making (and unmaking)."[43] Is this sense of "world" counterproductive for the existential effort that Rose is articulating? Does it undermine her urgently important effort to care for intimate alterities in the midst of ecological emergency? Must ecological existentialism get over the world, or is there still more world to come?

Rose notes that her sense of "world making" takes to heart the concept of world in Hannah Arendt's work and Haraway's concept of "worlding."[44] Although it is anthropocentric in orientation and it presupposes a phenomenological sense of world as horizon, Arendt's understanding of the human condition in its orientation toward Earth and world is nonetheless profoundly significant for any attempt to consider a new beginning for planetary coexistence.[45] While Haraway's concept of worlding is definitely not anthropocentric, and it does not think of world in terms of a phenomenological horizon, her sense of companion species as "full partners in worlding, in becoming with" does seem to reduce specific beings to their constitutive intra-actions, such that the significant otherness of companions loses its otherness.[46] Things have constituent parts, which are also things, but no single scale of existence is reducible to another scale. Atoms are not more real than molecules, and my body is not more real than my heart. Each thing is irreducibly itself. Furthermore, beings are not static or eternal substances. They change, and what came before changed, and so will what comes after them, but they are not reducible to their processes or relationships, nor can they be said to emerge from processes or relationships.

Things emerge from other things, with no homogeneous "nature" or process-relational "matter" underlying them. Accordingly emergence must be understood as a trait of things. That runs contrary to typical materialistic and naturalistic interpretations of emergence, which consider things to be abstractions or effects of self-organizing dynamics, plastic and vibrant materialities, or dynamic becomings, which undermine alterities by dissolving them in a relational solution. Emergence is simply another term for nature, *phusis*, the upwelling of being, coming to presence. It is not the reality of the myriad things but their manifestation in contact with one another. "Whether you call it *emergence* or *appearance*," it is not the irremediable alterity of strange strangers but is their mediated contact in the mesh.[47] By undermining the specificity of intimate alterities, materialist and naturalist theories of emergence fail to let the current ecological emergency emerge.

On one hand, for ecological existentialism to do justice to intimate alterities, it must orient itself toward an ecology without matter, world, or nature. That means affirming our utter *Weltlosigkeit*. It means dropping the world.[48] On the other hand, although it maintains tenuous alliances with proliferating postal movements (postmodern, postcolonial, postsecular, postnational, postnatural, postfeminist, posthumanist, postmetaphysical, posthistorical, not to mention post-post-whatever), ecological existentialism can affirm a more complex sense of being "after" the world, one that avoids the linear relays of the post as well as its transcendence and triumphalism. We are after the world, not only subsequent to the world—good riddance to hegemonic homogeneity!—but also seeking it, attending to its arrival in the heaping, swarming, piling up of countless others. In this case, Morton's announcement of the end of the world seems hyperbolic, or worse, it resonates with triumphalist tones, like Jesus in the Gospel of John: "I have overcome the world" (John 16:33). To be sure, this does not resonate with that more apocalyptic John, whose Book of Revelation is the final book of the Bible in sequence and in topic, envisioning an extremely violent end to all of existence, in which only a select few humans receive salvation while the remaining multitudes are consigned to a lake of fire. Such apocalypticism is exactly what Morton would like to drop, as "the end of the world is the end of endings," which is to say, the end of fantasies about total dissolution, annihilation, cataclysmic purification.[49]

There is no apocalyptic either/or morality in Morton's worldlessness. He maintains vigilant attention to the dangers of the beautiful soul syndrome that would divide people into those ("us," or at least "me") who understand the worldlessness and emptiness of things, and the bad or evil group over there ("them"), where things are reduced to worlding or world making. If the end of the world is an end of endings, it is not an absolutely final end. "This end is precisely not an instant vaporization, but rather a lingering coexistence with strange strangers."[50] If coexistentialism is an obstetric posthumanism, it is facilitating the birth of ecology out of the spirit of the end of the world: "coexistence—that is after all what ecology profoundly means."[51] This affirmation of things coexisting without world does not preclude a concept of world, that is, as long as the world or worlding is not an underlying process, supersized container, or phenomenological horizon but is a trait of things themselves or, at the very least, a rhetorical term of solidarity that can be deployed pragmatically but does not refer to anything real.

Consider Morton's use of a worldly term like "Universe" in his proposal for "*an object-oriented feminist ecology*," which joins the struggle of ecofeminism to understand and affirm the continuity (mesh) and difference (alterity) of humans and nonhumans: "A Universe of unique beings would benefit ecological feminism."[52] On one hand, the word "Universe" is a term of solidarity with queer theorists and ecofeminists, particularly in light of the challenges that Morton is posing to the intra-active undermining of things in Karen Barad's *Meeting the Universe Halfway*.[53] By speaking of a Universe, Morton is meeting Barad halfway.[54] Terms like "world," "cosmos," and "universe" are not banned from theoretical discourse or everyday speech. On a trivial level, one could multiply examples of Morton using the word "world" in conventional statements like "in the real world."[55] On the other hand, the notion of a Universe of unique beings is not only a rhetorical strategy for building solidarity across different movements (new materialism, queer theory, ecofeminism, object-oriented ontology, process-relational philosophy). It is indicative of the possibility of conceiving of worldhood not as a background, container, horizon, material relationship, process of becoming, or sum total of beings, but as the coexistence of exuberant entities, shimmering singularities piling, amassing, assembling, stretching out, extending, without a background, organization, foundation, or structure.

Nancy considers the arrival of this kind of worlding (*mondialisation*) after the end of the world, a worlding that is the dispersion of beings, piling up, the heap, *partes extra partes*, without a foundation.[56] For Nancy as for Haraway, *mondialisation* is an alternative to globalization (*globalisation*), conveying a possible opening in contrast to neoliberal enclosures. Mobilizing species and their significant otherness, Haraway mentions that practices of "becoming with" companion species are practices of "becoming worldly" in the sense of an "autre-mondialisation," which is an alternative to "militarized neoliberal models of world building" and an alternative to the reactionary antiglobalization movement.[57] Worlding in that sense is "cosmopolitics," which is "a practice for going on, for remaining exposed to consequences."[58] Cosmopolitics is an alluring proposition that provokes coexistential training—"learning to play with strangers"—which in the context of human-dog relations means "that people must learn to meet dogs as strangers, as significant others."[59] This is different from the humanism of classical and modern cosmopolitanisms. It is more like "eco-cosmopolitanism," a concept developed in the ecocriticism of Ursula Heise. "Eco-cosmopolitanism, then, is an attempt to envision individuals and groups as part of planetary 'imagined communities' of both human and nonhuman kinds."[60]

In an eco-cosmopolitan sense, worlding is indicative of ethico-political concern for the agency and alterity of all beings, human and nonhuman. Ella Myers gives a compelling account of such an ethics and politics of "worldly things," drawing on three thinkers in the "shared intellectual heritage" of "existential phenomenology," Arendt, Levinas, and Foucault, who provide Myers with concepts of care for the world (Arendt), for the other (Levinas), and for the self (Foucault).[61] To clarify what she means by "worldly things," Myers draws on Bennet's concept of the vibrant "thing-power" of matter and Latour's concept of actors, including Latour's proposal for a politics of things (*Dingpolitik*), "object-oriented democracy."[62] In the case of cosmopolitics and object-oriented democracy, worlding is less about undermining things and more about bringing nonhumans into ethical concern and opening up the struggle for the collectivization of humans and nonhumans. For Myers, following Morton's use of the term, this worldly concern can be called "coexistentialism."[63]

However, in its ontological register, worlding nonetheless tends to undermine the irreducible specificity of beings. Even Latour's actors and Bennett's thing-power threaten to undermine things. Are actors merely their actions? Is thing-power an effect of matter? However quantized, heterogeneous, vibrant, and dynamic action and matter might be, they nonetheless undermine strange strangers by defining them by their processes and relationships in the mesh. Nancy's articulation of worlding comes closer to an ecology of things. For Nancy, the opening of the world is not an intra-active process or a container, nor is it the opening of a phenomenological horizon. It is the spatiotemporal opening of each being, and every spatiotemporal unit is an integral island withdrawn from other islands, *partes extra partes*.

> That the world is there means that it is nowhere since it is the opening of space-time. That it is coexistence means that its opening opens it in all the senses, *partes extra partes*, spatio-temporal dis-positing dispersion, and between space and time just as the one in the other, a manner identical to its proper distention.[64]

Coexistence is this worlding which is not one. "The one-and-only-ordered-world is crumbling," as Nancy observes with the physicist Aurélien Barrau, but we are not simply left without a world.[65] The end of that world is giving way to at least two, perhaps the couple.

The opening of the world is the exposure of 1 + *n* objects, which always means that the world is always already two worlds, at the very least, as with the bipolar intimacy of Sloterdijk's microspheres. Recognizing the alterity of any other being renders untenable a single world. You do not have to speculate on worlds outside of the 13.8 billion-year-old mesh of strangers conventionally called the "universe" in order to see that the universe must be a pluriverse or multiverse, "unique yet nonunified, universal yet multiversal."[66] A glance at a neighbor will suffice. A recognition of the other is a recognition of this worlding which is not one. "As soon as I recognize the otherness of the other as irreducible to me or to my own, the world itself becomes irreducible to a single world: there are always at least two worlds."[67] The at least two (1 + *n*) worlds are incommensurable, irreducibly other, withdrawn from access, and as such, they can be conceived as islands more than worlds. Consider Derrida's remarks.

> Between my world, the "my world," what I call "my world"—and there is no other for me, as any other world is part of it—between my world and any other world there is first the space and the time of an infinite difference, an interruption that is incommensurable with all attempts to make a passage, a bridge, an isthmus, all attempts to communication, translation, trope, and transfer that the desire for a world or the want of a world, the begin wanting a world will try to pose, impose, propose, stabilize. There is no world, there are only islands. . . . We are *weltlos*.[68]

Derrida mentions here that this worldlessness can be interpreted in terms of a poem by Paul Celan, "Grosse, Glühende Wölbung" (Vast, Glowing Vault), the final line of which is, "Die Welt ist fort, ich muss dich tragen" ("The world is gone, I must carry you"). This carrying holds life and death. A mourner carries the dead, holding death in life, intimately bearing the loss of the other. A mother carries a child in her womb and in her arms, intimately bearing the arrival of the other. Without a world to carry everything and everyone, we carry one another.

Worldlessness opens onto the unbearable intimacy of ecological responsibility. I must carry you, a stranger whose arrival and departure displays an exorbitant nothingness, an irremediable alterity. Carrying you means holding within me that which is withdrawn from any access or appropriation, bearing within myself that which remains unbearable. Carrying you, I am under the weight of the world, your world, the world of the other, the world as the other: "The world after the end of the world."[69] This world without world is not a background, container, or horizon. It is the expansive singularity of every other, the opening of each being. This is not one opening for all beings. That sounds like a phenomenological horizon or a process-relational background. The opening of the world opens as each singularity. When a being comes into existence, the world begins, and when anything dies, when any singularity passes out of existence, that is the end of the world.

> For each time, and each time singularly, each time irreplaceably, each time infinitely, death is nothing less than an end of *the* world. Not *only one* end among others, the end of someone or of something *in the world*, the end of a life or of a living being. Death puts an end neither to someone in the world nor to *one* world among others. Death marks each time, each time in defiance of arithmetic, the absolute end of the one and only world, of that which each opens as a one and only world, the end of the unique world, the end of the totality of what is or can be presented as the origin of the world for any unique living being, be it human or not.[70]

Amid ecologies of others coming and going, humans have always been after the end of the world. The world has already ended, and it will continue ending, each time irreplaceably, in the loss of the other. The ending that coincided with the arrival of hyperobjects and the deaths of God and nature is a unique ending, as it put an end to the world behind the scenes, the world that confines or undermines the exuberant nothingness of things, the world that prevents us from carrying the world of every other. After the end of that world, it is possible once again and for the first time to carry the world, carrying each other and being carried away by each other. Dropping the world as background, horizon, and container, a mesh of strange strangers shows up, continually reopening the question of their worlding. Who is carrying who? How many worlds are there in this (these) crowd(s) of strangers? Amid the swarming singularities of coexistence, who counts?

> We do not conclude: We open up, we attempt. *Dans quel(s) monde(s) vivons-nous? [What's this (these) world(s) coming to?]*—the ambiguity of whether it is a pluralized singular or singularized plural is admitted—may be heard in the same key as: "Is there something like a world in which we find ourselves?" Or perhaps: "Of all the possible worlds, which one is ours?" Or might we even understand the "world" in the sense of when one cries out: "The whole world is here!" By which crowd, which pack, which fray are we being carried away? Can one even identify, distinguish, and disentangle such a web?
>
> Whatever the case may be, today, one must get in the fray.[71]

NOTES

1. Derrida, *The Politics of Friendship*, 1–2.
2. Ibid., 22.
3. Morton, "They Are Here," 184.
4. Nancy, *Sense of the World*, 4. For a philosophical history of the modern sense of the world, including the deconstructive end of the world and the possibility of philosophy without world, see Sean Gaston, *The Concept of World from Kant to Derrida* (London: Rowman & Littlefield International, 2013).
5. Morton, "Deconstruction and/as Ecology," 301.
6. Jean-Luc Nancy and Aurélien Barrau, *What's These Worlds Coming To?*, trans. Travis Holloway and Flor Méchain (New York: Fordham University Press, 2015), 48.
7. Morton, *Realist Magic*, 58.
8. Luce Irigaray, *Elemental Passions*, trans. Joanne Collie and Judith Still (New York: Routledge, 1992), 55.
9. Morton, *Hyperobjects*, 7.
10. Ibid., 126.
11. Nietzsche, *Twilight of the Idols*, 24.
12. Thacker distinguishes the alterity of the planet from subjective "*World*" and objective "*Earth*." Thacker, *In the Dust of This Planet*, 6. On nature-culture dualisms framed as Earth-world dualisms in environmental thinking, consider the following: Johan Rockström and Mattias Klum, *Big World, Small Planet: Abundance within Planetary Boundaries* (New Haven: Yale University Press, 2015); Clifford Chalmers Cain, ed., *Many Heavens, One Earth: Readings on Religion and the Environment* (Lanham: Lexington Books, 2012). For an interpretation of Earth-world couplets in Kant, Arendt, Heidegger, and Derrida, see Kelly Oliver, *Earth and World: Philosophy after the Apollo Missions* (New York: Columbia University Press, 2015).
13. Hans Jonas, *The Imperative of Responsibility: In Search of an Ethics for the Technological Age*, trans. by Hans Jonas with David Herr (Chicago: University of Chicago Press, 1984), 6–7.
14. Husserl, *The Crisis of the European Sciences and Transcendental Phenomenology*, 142–43.
15. Morton, *Hyperobjects*, 76.
16. Emmanuel Levinas, *Existence and Existents*, trans. Alphonso Lingis (Dordrecht: Kluwer, 1988), 52–64.
17. Ibid., 13–15.
18. Graham Harman, *Towards Speculative Realism: Essays and Lectures* (Winchester: Zero Books, 2010), 93. Morton is explicit that his form of object-oriented ontology follows Graham Harman's object-oriented philosophy. See Morton, *Realist Magic*, 9, 19.
19. Graham Harman, "On the Undermining of Objects: Grant, Bruno, and Radical Philosophy," in *The Speculative Turn: Continental Materialism and Realism*, ed. Levi Bryant, Nick Srnicek, and Graham Harman (Melbourne: re.press, 2011), 24.
20. Karen Barad, *Meeting the Universe Halfway: Quantum Physics and the Entanglement of Matter and Meaning* (Durham: Duke University Press, 2007).
21. Haraway, *When Species Meet*, 165.
22. Harman rejects the notions that objects must be natural and the additional criteria of Leibniz that substances must be simple and indestructible. Harman's objects can be artificial, composite, and destructible, like a bridge, a refrigerator, or a multinational corporation. Graham Harman, *The Quadruple Object* (Winchester: Zero Books, 2010), 18.
23. Heidegger, *Being and Time*, 99.
24. On the parallel between Morton's mesh and strange strangers and Harman's sensual objects (related) and real objects (withdrawn), see Morton, *Hyperobjects*, 118.
25. Harman, *Towards Speculative Realism*, 124.

26. Harman, *The Quadruple Object*, 70–72. Al-Ghazalī is perhaps the most well-known proponent of occasionalism in Islamic theology. Abū Hāmid Al-Ghazalī, *The Incoherence of the Philosophers*, 2nd ed. (Provo: Brigham Young University Press, 2000).

27. Catarina Belo, *Chance and Determinism in Avicenna and Averroes* (Leiden: Brill, 2007), 213.

28. Morton, *Realist Magic*, 73–74.

29. Keiji Nishitani, *Religion and Nothingness*, trans. Jan van Bragt (Berkeley: University of California Press, 1983), 4.

30. As Jay Garfield notes in his commentary on Nagarjuna's *Mūlamadhyamakakārikā*, "emptiness" is a misleading translation: "The emptiness that is referred to by *sunyata* is the real universe as it is when it is empty of our preconception." Nagarjuna, *The Fundamental Wisdom of the Middle Way* , 256.

31. Nishitani, *Religion and Nothingness*, 93. The Latin *sive* (literally meaning "or") is a translation of the Japanese word *soku*, which Nishitani employs in a way similar to traditional philosophical uses of the Latin *qua*, indicating that one term is an alternative to the other or in reciprocal alternation with the other. For Nishitani, it indicates the reciprocity of the terms involved (ibid., 303).

32. Ibid., 75.

33. *The Heart Sutra: The Womb of Buddhas*, trans. and comm. Red Pine (Berkeley: Counterpoint, 2004), 2.

34. Timothy Morton, "Buddhaphobia: Nothingness and the Fear of Things," in *Nothing: Three Inquiries in Buddhism*, Marcus Boon, Eric Cazdyn, and Timothy Morton (Chicago: University of Chicago Press, 2015), 222, 231; cf. Morton, *Realist Magic*, 47; Morton, "Elementality," 273–74. For Tillich's account of this distinction, see Paul Tillich, *Systematic Theology: Volume 1* (Chicago: University of Chicago Press, 1951), 188.

35. Morton, *Realist Magic*, 47.

36. Morton, "Buddhaphobia," 201.

37. Ian Bogost, *Alien Phenomenology, or What It's Like to Be a Thing* (Minneapolis: University of Minnesota Press, 2012). It is important to distinguish Bogost's alien phenomenology from that developed by Bernhard Waldenfels. Reminiscent of Husserl, Schutz, Merleau-Ponty, and Levinas, Waldenfels' *Phenomenology of the Alien* explores the place of otherness in human subjectivity, including the otherness given in one's experience of one's own body and in experiences of the corporeality of other people. However, Waldenfels does not address nonhuman aliens. Bernhard Waldenfels, *Phenomenology of the Alien: Basic Concepts*, trans. Tanja Stahler and Alexander Kozin (Evanston: Northwestern University Press, 2011).

38. Bogost, *Alien Phenomenology*, 85.

39. Ibid., 79.

40. Ibid., 78.

41. Morton, *Realist Magic*, 68.

42. Rose, *Wild Dog Dreaming*, 51.

43. Ibid., 11–12.

44. Ibid., 147n27.

45. See chapter 11, "Indications of an Axial Age."

46. Haraway, *When Species Meet*, 301.

47. Morton, *Hyperobjects*, 119. For more on an object-oriented account of emergence, for which emergence is an effect of things, not the other way around, see Morton, *Realist Magic*, 136–44.

48. Morton suggests "dropping Nature, *world*, and so on," and he elsewhere muses that "to drop the concepts Nature and *world*" would be "the ultimate environmentalist argument." Morton, *Hyperobjects*, 100, 121.

49. Ibid., 95.

50. Ibid.

51. Ibid., 128.

52. Morton, "Treating Objects like Women," 51, 67.

53. Ibid., 60.

54. Along with Barad's universe, Morton's universe also resonates with the universe of another process-relational philosopher, Steven Shaviro, who draws on Whitehead in an attempt to meet the universe of object-oriented ontology halfway. Steven Shaviro, *The Universe of Things: On Speculative Realism* (Minneapolis: University of Minnesota Press, 2014).
55. Morton, *Realist Magic*, 29, 168.
56. Jean-Luc Nancy, *The Creation of the World* or *Globalization*, trans. François Raffoul and David Pettigrew (Albany: SUNY Press, 2007), 27–28.
57. Haraway, *When Species Meet*, 3.
58. Ibid., 106. Haraway adapts the term "cosmopolitcs" from Isabelle Stengers. I elaborate on this sense of cosmopolitics without world elsewhere. See Sam Mickey, *On the Verge of a Planetary Civilization: A Philosophy of Integral Ecology* (London: Rowman & Littlefield International, 2014), 183–92.
59. Haraway, *When Species Meet*, 243.
60. Ursula K. Heise, *Sense of Place, Sense of Planet: The Environmental Imagination of the Global* (Oxford: Oxford University Press, 2008), 61.
61. Myers, *Worldly Ethics*, 2, 14.
62. Ibid., 97–103.
63. Ibid., 180n55. Aware that Morton "rejects the terminology of world," Myers maintains the term, considering Morton's "*mesh*" as "a rough synonym for world" in her sense (ibid., 187n63). Whether that synonym is too rough to be felicitous is an open question. Her definition of coexistentialism as "deep interconnectedness" and her omission of Morton's concept of strange strangers suggest that her coexistentialism does not adequately address the uncanniness and uncertainty always already interrupting interconnectedness (ibid., 101). While she discusses alterity, it is mostly restricted to humanistic limits (ibid., 55–56, 60, 66, 78).
64. Jean-Luc Nancy, *The Creation of the World*, 73.
65. Nancy and Barrau, *What's These Worlds Coming To?*, 5.
66. Ibid., 2. On the tangled history of philosophical, scientific, and religious perspectives on theories of multiple worlds, see Mary-Jane Rubenstein, *Worlds without End: The Many Lives of the Multiverse* (New York: Columbia University Press, 2014).
67. Luce Irigaray, *Sharing the World*, trans. Stephen Pluhacek (London: Continuum, 2008), x.
68. Jacques Derrida, *The Beast and the Sovereign, Volume II*, trans. Geoffrey Bennington (Chicago: University of Chicago Press, 2011), 9.
69. Jacques Derrida, "Rams: Uninterrupted Dialogue—Between Two Infinities, the Poem," trans. Thomas Dutoit and Philippe Romanski, in Jacques Derrida, *Sovereignties in Question: The Poetics of Paul Celan*, ed. Thomas Dutoit and Outi Pasanen (New York: Fordham University Press, 2005), 140. Derrida also elaborates on Celan's "Grosse, Glühende Wölbung" in this essay (ibid., 141).
70. Ibid., 140.
71. Nancy and Barrau, *What's These Worlds Coming To?*, 7.

TEN

Askesis: Shut Up and Train!

It is becoming horrifyingly apparent that human existence has woven itself precariously into the geological and biological systems of Earth, not only locally, but on a planetary scale, as is evident in human impacts on global biodiversity and the chemistry of the atmosphere and oceans. Having become vulnerable to the consequences of the exceptionally hypertrophic humans of modernity, humans today are beginning, perhaps, to learn to undo their fantasies of omnipotence and omniscience, beginning to welcome something different. This beginning did not irrupt out of nothing. It is a protracted beginning that can be seen emerging on a large scale during the twentieth century, with the development of a variety of emancipatory philosophical and social movements oriented toward critiquing the dominant systems of human civilization, e.g., critical theory, environmentalism, feminism, the civil rights movement, postmodernism, postcolonialism, and queer theory. While the ongoing contributions of those movements to facilitate various kinds of freedom from oppression, their critical stance has limits. As Latour puts it, critique has run out of steam.[1]

For the most part, but not exclusively, those twentieth-century movements are oriented toward addressing problems and pathological symptoms, unmasking oppressive power dynamics and hegemonic narratives. There are not many celebratory or visionary threads woven into those movements, and in that sense, they resemble twentieth-century theories of language, which are also oriented toward diagnosing and critiquing problems. As Sloterdijk observes, the majority of twentieth-century theories of language—"ideology critique or analytic philosophy, discourse theory or psychoanalysis, a theory of the encounter or deconstruction"—understand language as "a medium of lack and distortion."[2] Language and discourse show up as "symptoms and problems," and are rarely interpreted as "vectors of affirmations and prophecies."[3] Sloterdijk draws on Nietzsche to recover a more festive sense of language, one of joyful affirmation, a celebration of existence.

Affirmation in the Anthropocene is not merely a linguistic endeavor. It is about overcoming the cynicism that sees everywhere only problems and symptoms of domination. There surely is domination, but that does not exhaust the reality of things. Affirmation is about overcoming beautiful soul syndrome, which disengages from practices of becoming worldly and erects an either/or morality that opposes "us" and "them"—resenting society ("them") for not conforming to the ideals of harmony and purity that the beautiful soul prefers. Fearing and loathing problems "out there" (in language, politics, society, or the environment) pushes those problems away as objects of criticism or moral judgment, rather than facing them and taking responsibility for their urgent demands. Cynicism fails to let the emergency of coexistence emerge. Countering that cynicism with affirmation does not mean abandoning critique for acquiescence and simply letting violence and exploitation run free. Affirmation is not passive acceptance of the violence and exploitation spreading through planetary coexistence, but neither does it fall into the cynical attitude of reactionary opposition, which still measures itself in terms of the violence and exploitation it opposes. Neither passive nor reactive, affirming coexistence is about exercise (*askesis*). It is about practicing—acting and being acted on differently, remaining exposed in intimate entanglements, becoming worldly by becoming with others.

Practice is not perfect, although it might promise to make perfect. It is called "practice," not "achievement." Improving or intensifying capacities through training means doing things that you are not yet able to do, being something that you are not, faking it as a way of making it. That inconsistency is not a bad thing. It is indicative of the rift between the irreducible specificity of things as intact beings and their appearance in mediated contact with other beings. The challenge of being human amid intimate alterities demands sincere engagements in the hypocrisy of practice. Training thus provides an antidote to contemporary cynicism, allergy medicine for the beautiful souls who cannot tolerate anything that contaminates their purity and harmony. There are countless practices that can be taken up, recovered from traditions or invented anew. For Sloterdijk, the philosophical asceticism of the ancients is an exemplary antidote to contemporary cynicism. A recovery of ancient Cynicism (Sloterdijk's "kynicism") counters the critical function of modern cynicism with practices of existential self-affirmation.[4] However, this existentialist kynicism should not be confused with an accurate historical account of the practices of actual Cynics, who may not have emphasized self-affirmation as much as Sloterdijk suggests.[5] The point here is not about the specific exercises of Cynics as much as it is about the importance of taking seriously the challenge of abandoning the attitude of resentful criticism and practicing affirmative self-transformation. In short, "complaint-addicted moderns" should stop complaining and start training.[6]

While training might sound more conservative than progressive, more Right than Left, it need not be. Consider Žižek's radical leftist perspective. "In today's era of hedonist permissivity which serves as the dominant ideology, the time has come for the Left to (re)appropriate discipline and the spirit of sacrifice: there is nothing inherently 'fascist' about these values."[7] Žižek is commenting here on a quotation from an unpublished 2007 interview of his friend Alain Badiou, in which Badiou calls for a popular discipline, saying that "those who have nothing have only their discipline." The poor, those with no financial or military means, those with no power, all they have is their discipline, their capacity to act and be acted on together. This discipline is already a form of organization, becoming-with and becoming worldly. Practice is already a change.

EXERCISING

If there is a bumper sticker for coexistentialism, it is probably not the dubiously harmonious "coexist" sticker. It is more likely to be found in something closer to the "Shut Up and Train" bumper sticker, found on the vehicles of gym-lovers and dog-lovers alike.[8] For Haraway, that phrase is a possible slogan for her work with companion species, although, in the end, she prefers a slogan that makes sense to those who have already shut up and done some training with their significant others.

> Many pithy slogans might urge us on in trying to learn more about how to flourish together in difference without the telos of a final peace. A rough one from the dog world might be, "Shut up and train!" But I prefer to end with a longing that it might be said of me someday what good agility players say of those whose runs they admire, "She has met her dog."[9]

Affirming the finite conditions of planetary coexistence is not about reckless abandon, and it is not about resentful limitation and critique. It is about exercises in becoming-with and becoming worldly, meeting each other, meeting the challenges of living and dying together by practicing. For Haraway, there is no alternative to practice. Sloterdijk likewise understands the necessity of practice. "Anyone on earth who understands their situation will face the fact that no one leaves this place alive. People on this gloomy orb must practice."[10] Neither reckless nor reactionary, practicing means actively paying attention to the real in all of its roughness, absurdity, and unruly complexity. In other words, it means changing oneself by attuning to the exigencies of coexistence, which requires dropping the fantasies that maintain ideals of smoothness, harmony, purity, and innocence.

Practicing is not just about learning a particular skill but, in a more general sense, it is about humans designing and transforming themselves. In terms of a "General Ascetology," *askesis* is fundamental to human existence, such that human existence has "an autoplastic constitution," shaping itself in an "operationally curved space" (i.e., spheres), and the intimacy of that space allows your thoughts, feelings, and actions to reflect back on you, further changing your thinking, feeling, and acting, opening "self-referential relationships" whereby you can and must participate in the design of your own subjectivity.[11] In other words, human existence is its own spherological or immunological design project. "*Dasein ist design*."[12] You could say that this involves a postsecular integration of religious and secular exercises and training regimes. However, Sloterdijk argues that this generalized *askesis* undoes the distinction between that which is religious and that which is not. In short, "no 'religion' or 'religions' exist"; "The false dichotomy of believers and unbelievers becomes obsolete and is replaced by the distinction between the practising and the untrained, or those who train differently."[13] Zen meditation and Catholic prayer are not different in kind from farming, surfing, kayaking, chemistry, or football. They are all training regimes, different to be sure, but with no great religion/secular divide. Each training regime offers different possibilities for individuals and groups to design themselves, to transform themselves in light of "vertical tensions," which mark intensified capacities for acting and being acted upon.[14]

The term "anthropotechnics" can be misleading. Like "Anthropocene," it might sound "anthropocentric," although it is naming an openness that undoes any pretense of centrality. The value of the human is plastic, open to design, a design without design, and regardless of the specificities of the design, humans can never separate themselves from the nonhumans that figure into the conditions of sphere construction. Being in intimate immune systems of spheres, no center can escape intimate contact with peripheries. Furthermore, the call to participate in anthropotechnical exercises does not come from inside, not from within the human. The title of Sloterdijk's book on anthropotechnics is not *I Must Change My Life*. The call comes from the outside. The call to exercise is exemplified in Rilke's poem, "Archaic Torso of Apollo," in which the subject of the poem is observing a sculpture of a torso and, feeling seen by the torso ("there is no place / that does not see you"), concludes with these words, "You must change your life."[15] Being seen is not about reciprocity of seer and seen—me observing and being observed by the torso. As Sloterdijk observes, it is a reversal of the seer-seen dynamic, such that the torso "eyes me more sharply than I can look at it."[16] It grasps me when it sees me, compelling me to change my life. Meanwhile, I am rendered incapable of seeing the torso, blinded by its provocative gaze, by the "glow" and "glisten" of the torso. "So we are grasped," as Rilke says toward the middle of "A Walk," "by what we cannot grasp."[17]

After God, nature, human, and world, humans are grasped by the other, remaining exposed to the outside. This "thought of the outside" is indicative of the spherological catastrophe of modernity, and it finds profound affirmation in "the authors of existentialism," who opened the way for a "second wave" of the thought of the outside in "post-existentialist" philosophy, which "Foucault set the tone for."[18] In the fashion of coexistentialism, being out is in. "The outside is the most proximate, the innermost and the native, and everything inside is merely a molding or folding of the outside."[19] Anthropotechnics is not anthropocentric. It is about practicing eccentric sphere construction. Anthropocentric sphere construction is an autoimmune disease, protecting an exclusively human center and marginalizing nonhumans, thus destroying the spheres in and as which humans exist, like sawing off the limb you are sitting on.

The untenability of anthropocentric sphere construction is especially true in the Anthropocene, in which humans and nonhumans are woven together at a planetary scale. In the climate emergency marked by the deaths of God and nature, the humanism of Enlightenment rationality is giving way to a convergence of anthropotechnics and "atmotechnics," wherein humans play an active role in shaping the atmosphere while remaining horrifyingly far from being in control of the situation: "Humans create their own climate—not of their own free will, however, under self-chosen circumstances, but under found, given and handed-down ones. The tradition of all dead climates burdens the moods of the living like a nightmare."[20] Insofar as anthropotechnics is situated in the Anthropocene, anthropotechnics is intimately intertwined with practices for understanding and responding to anthropogenic impacts on the planet's ecosystems. Anthropotechnics is thus aptly described with Nancy's term, *ecotechnics*. It is the art of ecology without nature: "*ecotechnology* where any kind of 'nature' develops for us (and by us)."[21] When "the *oikos* is everywhere," any sovereignty (human, law, God, or nature) is disturbed, weakened, damaged: "ecotechnics washes out or dissolves sovereignty."[22] On one hand, the dissolution of sovereignty opens onto the nihilism of economic globalization, which forgoes cooperative solidarity for a war of foams, and the nihilism of the cynic, who retreats into a pure interiority from which to criticize the evil world. On the other hand, ecotechnics opens up possibilities for the development of cooperative asceticisms for planetary survival, avoiding hegemonic sovereignties and their nihilistic leftovers.

The human condition in the Anthropocene calls for anthropotechnics exercises on a planetary scale, since the condition of humans anywhere is entangled in the mesh with the conditions of humans and nonhumans around the planet, forcing humans to remain in an "overtaxing-field of enormous probabilities," in which "I am to estimate the effects of my actions on the ecology of the global society. . . . I am to develop into a fakir [ascetic] of coexistence with everyone and everything, and reduce my footprint in the environment to the trail of a feather."[23] Ecological emergency thus pushes anthropotechnics to its limit, forcing the development of a "global co-immunity structure," an immune system that wards off the foams of individualism so as to protect the intimacy constitutive of planetary coexistence.[24] Such a planetary immune system—a "co-immunism"— requires "universal co-operative asceticisms," and these "monastic rules must be drawn up now or never," and people must live by them, which means taking on "the good habits of shared survival in daily exercises."[25]

I discuss some examples of such exercises elsewhere in this book, including practices of heartfelt companioning with dogs (chapter 4), glancing at faces and surfaces (chapter 5), techniques of anthropocosmic ecstasy in poetry, religion, and phenomenology (chapter 7), and feminist, queer, prismatic, ecological, and psychedelic ways of looking good at and for the other (chapter 8). As a form of speculative reflection, object-oriented ontology (chapter 9) is also a practice, not just a theory. Attuning to things themselves, object-oriented ontology can be thought of as a radically non-anthropocentric phenomenology. As Husserl understood, phenomenology is a life-changing exercise. Specifically, it is a practice of suspense or bracketing (*epoché*), holding all preconceptions in abeyance, attending to what appears in consciousness while abstaining from all judgment. It will change your life. "Through the abstention which inhibits this whole hitherto unbroken way of life a complete transformation of all of life is attained, a thoroughly new way of life."[26]

For practitioners of object-oriented ontology, that new way of life is an uncanny life as a unique being in a universe of unique beings, including massively distributed beings—hyperobjects—like climate change, capitalism, the Internet, *Homo sapiens*, plastic, and black holes. Object-oriented reflection is a contemplative exercise in accepting things in all of their strangeness and connectedness. "Philosophical reflection on the hyperobject is also a form of rest. There are various dimensions of this rest—mindfulness, awareness, simple letting-be—that are equally fascinating. Rest in this 'positive' sense suggests a deep acceptance of coexistence."[27] To practice restful acceptance is not only to exercise, but also to exorcise. Affirming coexistence requires exorcising the ghost of the "nature" that already died along with God and so many grounds and backgrounds. Exorcising nature means contemplating the remains of planetary coexistence—"the charnel ground after the end of the world."[28] "Exorcise is good for you," as Morton observes: "Our continued survival, and therefore the survival of the planet we're now dominating beyond all doubt, depends on our thinking past Nature."[29]

Drawing variously from philosophy, religion, art, science, politics, technology, and practices of everyday life, daily exercises for shared survival are called for now more than ever in human history. Those practices can and must be taken up in many different ways, yet their differences cannot be disassociated. They must be solidarity practices, cooperative asceticisms, sharing a common thread: cultivating attention to the continuity and difference of all beings in the mesh of coexistence, human, nonhuman, and otherwise. In that sense, these cooperative asceticisms are varieties of contemplative ecology, wherein exercises in "attention" (*prosoche*) engage vertical tensions, training for "heightened awareness."[30] The vertical tensions of contemplative ecology intensify capacities for remaining vulnerable with others, capacities for carrying and being carried away by others, bearing the unbearable intimacy of planetary coexistence. To elaborate on a practice of attuning to the exigencies of integral entities, attending to the call to change your life, I consider how a practice of listening can become a coexistential exercise, following the listening practices of two perhaps unlikely companions, Aldo Leopold and Jean-Luc Nancy.

LISTENING

Listening is an important part of ecological thought, as is evident in the compelling power of Rachel Carson's warning about a spring without birdsong and in the varieties of resonant phenomena studied in the science of acoustic ecology.[31] Like the glance, listening can be thought of as a contemplative ecological practice, a kind of *prosoche*. One can hear Aldo Leopold approach listening this way in his classic of environmental literature, *A Sand County Almanac and Sketches Here and There*. For Leopold, listening is about attuning to the land. The land does not just make sound. It is having conversations and conferences. Leopold remarks that, listening to geese during warm April nights, "we love to listen to the proceedings of the convention in the marsh."[32] Those are sounds of politics and commerce, dividing geese while uniting them in shared struggles. Leopold says that the geese have had a sense of unity long before twentieth-century notions of "the unity of nations": "Every March since the Pleistocene, the geese have honked unity," and through "this international commerce of geese," Leopold observes, "the whole continent receives as net profit a wild poem dropped from the murky skies upon the muds of March."[33]

For Leopold, listening is a scientific practice, like the way you can "listen for the final proof of spring" if you "cock your ears at the sky, dial out the bedlam of meadowlarks and redwings," and wait for the "flight-song of the upland plover," a migrating bird whose return from Argentina marks the beginning of springtime.[34] Partly meditation and partly science, listening can facilitate participation in the complex negotiations that mark seasonal changes or territorial changes, as when listening to birds proclaim the boundaries of their territories. Listening harbors the challenge of vigilantly awaiting the arrival of the unpredictable and rare. "There is a peculiar virtue in the music of elusive birds," which makes it worth staying up late to learn "the lessons of the night" or getting up early in the morning, maybe too early: "To arrive too early in the marsh is an adventure in pure listening; the ear roams at will among the noises of the night, without let or hindrance from hand or eye."[35] These diurnal and nocturnal lessons convey the humble, hyposubjective disposition of restraint that accompanies listening exercises.

It is not only animal life that sings. Leopold refers to singing waters that can be heard in a raging river or a babbling brook, and even to singing stone, as when lightning on a mountain causes explosions of "smoking slivers of stone that sing past you" and splinters that fly off of pine trees, sometimes stabbing into the ground and "humming like a tuning fork."[36] Artificial objects also have their own musical quality. For Leopold, "there is music in the sharp plane, the sharp chisel, and the sharp scalpel."[37] Leopold observes that, when a marshland was drained for farming, "the song of the power shovel came near being an elegy" for the cranes, endangering the "crane-music" and "the listening land."[38] In times of distress and destruction, the vigilant attention of listening can verge on apocalypticism, as does Leopold when he prophesies that, eventually, "the last crane will trumpet his farewell. . . . High out of the clouds will fall the sound of hunting horns, the baying of the phantom pack, the tinkle of little bells, and then a silence never to be broken, unless perchance in some far pasture of the Milky Way."[39]

In its mundane and apocalyptic moments, the land is not only resounding. It is also listening. Leopold describes his wonderment at the listening land in his encounter with the wind as it "makes music in November corn": "I sit and listen to the universal roar," and then there is "the honk of geese, invisible, but coming on," followed by a barking dog: "It is strange how the world cocks its ears at that sound, wondering."[40] Furthermore, not only is the land resounding and listening, it is a performing evolutionary orchestra. When describing how the slow speed of geological evolution fosters the growth of our appreciation for the value of the crane—a value exceeding the grasp of words—Leopold describes the call of the crane: "When we hear his call we hear no mere bird. We hear the trumpet in the orchestra of evolution."[41] Humans are called to tend to this resonant value of the crane and of all inhabitants of the land, all performers in this rather strange orchestra. This is the task of researchers and professors, who Leopold notes have "the duty of examining the construction of the plants, animals, and soils which are the instruments of the great orchestra."[42] However, specialists often overspecialize in taking apart one instrument in the orchestra for the sake of science and progress, such that many of the instruments, especially the intricate and delicate ones, are "broken in the rush to spread progress."[43]

Leopold does not elaborate on the role of dissonance and discord in this evolutionary orchestra. While this orchestra may sound like a vibrating mesh of strange strangers, it might also succumb to beautiful soul syndrome, hearing only harmony and no strangeness, "only reconcilable and federative differences, far removed from bloody struggles."[44] This syndrome seems to show up in Leopold's definition of conservation as "a state of harmony between men and land," with land being understood as a community.[45] However, there are at least two indications that Leopold is not suffering from beautiful soul syndrome: his recognition that an attentive practice of listening must listen indirectly, and his attribution of listening to a mountain. Although he is famous for the phrase that titles his essay, "Thinking Like a Mountain," Leopold suggests the importance for humans to think *and listen* like a mountain. The opening sentences imagine the echoing howl of the wolf: "A deep chesty bawl echoes from rimrock to rimrock, rolls down the mountain, and fades into the far blackness of the night. . . . Only the mountain has lived long enough to listen objectively to the howl of a wolf."[46] All living things heed this call, but the mountain listens at a different scale, tuning in at a hyperobjective frequency. The mountain listens for this howl explicitly and in its implications, its indirect resonances. The howl is "implicit in a hundred small events: the midnight whinny of a pack horse, the rattle of rolling rocks, the bound of a fleeing deer, the way shadows lie under the spruces."[47]

In the way a glance meets the occluding edges of a surface and takes in that which shows up as hidden, listening attunes to the implicit resonances of the other. Consider the song of a river, like the Rio Gavilan in Chihuahua, Mexico. Listening to the river, Leopold notes that a river's song "ordinarily means the tune that waters play on rock, root, and rapid," but it means the vitality of the river itself: "The life of every river sings its own song, but in most the song is long since marred by the discords of misuse."[48] "This song of the waters" is strange, like the howl of the implicit, resounding with music that is "by no means audible at all," inaudible yet overwhelmingly close, strange but unbearably intimate: "a vast pulsing harmony—its score inscribed on a thousand hills, its notes the lives and deaths of plants and animals, its rhythms spanning the seconds and the centuries."[49]

Leopold strikes a cynical tone in his comments on parks, which "are made to bring the music to the many, but by the time many are attuned to hear it there is little left but noise."[50] Previous cultures were more harmoniously attuned to the songs of rivers than the current culture of parks and recreation. Leopold claims that the misuse of rivers is tied to an ideal of the good life oriented toward technological and scientific development with no musicality. Leopold has doubts about this sense of the good life, suggesting instead that the good life has a lot to do experienceing its music and preserving some music so that it can continue to resonate for others. Preserving that music is a challenge of ongoing harmonizing in the vast pulsing orchestra of evolution, striving for a harmony that Leopold knows will not be attained. It remains to come. "We shall never achieve harmony with land, any more than we shall achieve justice or liberty for people . . . the important thing is not to achieve, but to strive":

> The problem, then, is how to bring about a striving for harmony with land among a people many of whom have forgotten there is any such thing as land, among whom education and culture have become almost synonymous with landlessness.[51]

To put it another way, bringing about striving for harmony with the land in an age of landlessness is like bringing about the struggle for macrospheric intimacy amid a war of foams, in which education and culture are almost synonymous with individualism. It is the challenge of practicing anthropotechnics in the Anthropocene. Shut up and train! In this case, that means shut up and listen! While Leopold's training in listening is of inestimable value for contemplative ecological exercises, the notes of beautiful soul syndrome are problematic. A more exposed and fragile listening is called for to attune to the intimate alterities of coexistence. For more intense training, one can follow the contemplative listening of Nancy.

For Nancy, the practice of listening approaches the resonant sense of the world: "To be listening is always to be on the edge of meaning, . . . a resonant meaning, a meaning whose sense is supposed to be found in resonance."[52] Nancy's articulation of listening revolves "around such a fundamental resonance, even around a resonance as a foundation, as a first or last profundity of 'sense' itself (or of truth)."[53] Existence is always already coexistence, and in the space of that "co-" is the possibility of a returning sense, a resonance. "The possibility of sense is identified with the possibility of resonance, or of sonority itself."[54] The idea of resonant meaning suggests that sound and meaning share something in common, namely, *renvoi* ("return" or "referral"). Referrals of meaning or sense move from person to person, from signs to things, from words to worlds, from states to qualities, from alterity to presence, etc. Referrals of sound are evident insofar as sound resounds, reverberates, and echoes: "it spreads in space, where it resounds while still resounding 'in me.'"[55] In short, "meaning and sound share the space of a referral."[56] The space of this referral, the space of this resonant meaning, is the opening of sense, where sense is not an otherworldly meaning or an underlying materiality but is the opening of the world. As Nancy says elsewhere, "the world *no longer has* a sense, but it *is* sense," which is to say, it is the circulation of sense in the opening of each being, *partes extra partes*.[57] Insofar as the referrals of sense are both modes of connection and differentiation, making contact while leaving intact, "sense is touching," but not in a way that privileges touch over other sensory modalities, for touching here names the contact and separation whereby the senses connect and differentiate.[58] To say that sense is touching does not preclude the truth of sense as resonance.

If you practice listening, the world is not a background, horizon, or underlying process of self-organization or intra-action. It is the resonant opening of the play between multiple senses of sense, traversing subjectivity (sensing, apprehension) and objectivity (sensed, apprehended) as well as intelligible, aesthetic, divinatory senses, and even the unheard-of sense of nonsense. The world is reverberating with sense(s), pulsing with a "vibrant spacing-out."[59] Sense, in this sense, is like a voice emerging from a mouth, a vibratory dispatch emitted from the open spacing of the body. To listen is to attend to this openness, this vibrant spacing-out of sense.

Nancy contrasts listening with hearing.[60] On the one hand, "to listen" (*écouter*) is to approach the spacing of the resonant sense of the world. On the other hand, "to hear" (*entendre*) implies "agreement" and "understanding" (*entente*) of sense. One who listens (*écoute*) is thus different than one who hears and understands (*entend*). Nancy states this distinction in the following way:

> If "to hear" is to understand the sense (either in the so-called figurative sense, or in the so-called proper sense: to hear a siren, a bird, or a drum is already each time to understand at least the rough outline of a situation, a context if not a text), to listen is to be straining toward a possible meaning, and consequently one that is not immediately accessible.[61]

Listening to the resonant spacing of sense is beyond a hearing or understanding of sense, and thus also "beyond agreement or harmony (*harmony* [entente] or *resolution* in the musical sense)."[62] In contrast to Leopold's sense of harmony as something not achieved but strived for amid discord, Nancy's account of listening is more affirmative of difference and discord, suggesting that sound "resounds always beyond a simple opposition between consonance and dissonance," for it is made of "*discordant harmony*," which is to say, sound is constituted by "an intimate harmony and disharmony among its parts."[63] Similar to Leopold, Nancy's practice of listening abjures anthropocentrism. Sense resonates far beyond human existence, calling humans to tune in to the evolving mesh of coexistence. The timbre (i.e., tone color) of the resounding world is "an evolving process, and, consequently, listening evolves along with it."[64]

Nancy and Leopold affirm the fundamental resonance of sense, such that sense is the vibrant spacing-out that composes the evolving timbre of the world. Sense is the resounding orchestra of evolution. Furthermore, Leopold and Nancy both affirm that listening to this resonant sense is an ethical practice. Thus, Nancy does not only say that the sense of the world is resonance. As a touching resonance, it is called to be tactful. Accordingly, Nancy also defines "*sense as ethos*."[65] The vibrant spacing-out of sense resounds with ethical repercussions. Ethics for Nancy is not simply a prescriptive schema or a system of rules, but is a matter of dwelling in the open space of freedom. The opening of freedom provides a "'shelter' of being *which is its ownmost ethos as the ethos or abode of the human being*."[66] This is similar to Heidegger's claim that "originary ethics" is not about value theory or systems of prescriptions and proscriptions, but is about the dwelling or abode of the human, that is, "the open region in which the human being dwells."[67] A key difference is that Nancy does not follow Heidegger's anthropocentric restriction of openness to human existence. All beings share in the resonance of sense, each with its own abode. For Nancy, *ethos* is "the opening of space, the spacious shelter of being in existence."[68] In other words, the *ethos* of the human is its dwelling in the vibrant space of sense. This spacious shelter makes room for the tactful togetherness of ethical contact. Spacing out and bringing together, *ethos* can be understood rhythmically. *Ethos*, for Nancy, "functions as something akin to 'rhythm,' 'cadence,' 'caesura,' or 'syncope' ('spacing,' 'pulsating')."[69] The *ethos* or abode of the human opens with syncopation, the rhythmic spacing of sense. This rhythm is a syncope or syncopation, "the beat of a blending together and a pulling apart," the beat of connecting and differentiating, presencing and withdrawing.[70] It is syncopated because there is always an excess of withdrawal, more alterity than could ever be displayed. The syncopated rhythm of sense dances with the evolving timbre of sense, the colorful tonality expressing the discordant harmony of singularities. "Rhythm, dancing, opens up timbre, which resounds in the rhythmed space."[71] Rhythm and timbre are different inflections of the vibrant spacing of sense. Ethical practice is a way of participating in syncopated rhythm, discordant harmony, and other inflections of the simultaneous blending (attaching) and separating (detaching) that opens the resonant space of sense.

The practice of listening is a practice of attuning to the sense of the world, attuning to the resounding coexistence of singularities, which means attuning to the syncopation and discord that compose the orchestra of evolution, which is to say, the rhythms and tones that compose the vibrant spacing of the sense of the world. Performing the "praxis of sense," listening participates in the "most original *ethics*."[72] Furthermore, Nancy's account of listening resonates with Leopold's insofar as they both suggest that ethical engagements with the vibrant spacing of sense facilitate participation in community. Leopold discusses ethics as "a kind of community instinct in-the-making."[73] While Leopold might have suffered from beautiful soul syndrome, hoping that the community of humans and the land would become a harmonious one, albeit a harmony that is never achieved but remains to come, Nancy's sense of community is more affirmative of the syncopation and discord that are always already interrupting harmonious relationships.

Unattainable, unworkable, and discordant, community for Nancy is at its best when it is "the inoperative community" (*la communauté désoeuvrée*).[74] The "with"—the "com-" (*cum*)—of the community is not a mere togetherness or simple unity, but is rather the pulsing opening that composes the discordant harmony of sense. The pulsing "with" of community is the resonating syncope, the syncopation of resonance. To say "we" is to participate in the discordant harmony of "with," the syncopated blending and separating that take place at the beating heart of community. It is not just that the harmony of community is still to come, still in-the-making, but that its harmony sounds good when it sounds discordant, just like looking good is at its best when it looks different. Saying "we" expresses unity, harmony, and resonance while pulsing with the discord and syncopation without solidarity becomes totalitarian solidity. Saying "we," for Nancy, "expresses 'our' being divided and entangled."[75] The practice of listening is a practice of preparing to say "we," a practice of attending to the resonance of sense as it divides and entangles, blends and separates, attaches and detaches, presences and withdraws. The "with" of community implies that saying "we" expresses continuity and difference between humans and nonhumans. "We would not be 'humans' if there were not 'dogs' and 'stones.'"[76]

Unlike Heidegger, Nancy does not rob from or foreclose on dogs and stones, leaving them poor and worldless. They are just as resonant as humans. If every being is vibrating with the discordant harmony of community, then each being is itself a sort of tuning fork (i.e., diapason), like the splinter of pine that Leopold saw "humming like a tuning fork" after the pine tree was struck by lightning.[77] Every self or subject is a tuning fork ("a diapason-subject") resonating with the vibrant sense of the world.[78] When one listens, one is approaching a subject, something "resonating from self to self."[79] Each body is "a reverberation chamber" or a "resonance chamber."[80] Along these lines, Nancy proposes a definition of "a resonant subject," such that the subject is an echo-chamber that resounds with the resonance of being, or, to be more exact, with "being as resonance."[81] This resonance echoes not only in the human voice, but also in the sounds that emerge "from an animal gullet or from any kind of instrument, even from the wind in the branches: the rustling toward which we strain or lend an ear."[82]

The resounding boundaries of the inoperative planetary community entangle and divide all inhabitants of the land and all performers in the evolutionary orchestra, such that we can "*say we for all being, that is, for every being, for all beings one by one, each time in the singular of their essential plural.*"[83] The creation of a planetary community of harmonious relationships between humans and the land requires cooperative practices of ecotechnics that resonate with the musical value of the world. The resonant values of coexistence cannot be assimilated into the representations of value in modern economics. In other words, the worlding (*mondialisa-*

tion) of the world cannot be assimilated into the homogenized totality of globalization. The creation of the world through *mondialisation* opens the closed world of economic value to the detuned tunes of intimate alterities. As "a space in which a certain tonality resonates," the opening of the world takes place "outside representation," and thus "the value of the world" is withdrawn, wholly other, outside of representations of use-value, exchange-value, or any economic value.[84]

Practices of listening are cooperative asceticisms for attuning to the resonant sense of planetary community after the end of the world, sounding good to and for others, carrying a tune and being carried away by the tunes of others. The value of the world after the end of the world is the very resonance of the pulsing rhythms and evolving timbre of sense. It is this resonance of beings themselves, strange tuning forks vibrating in irremediable alterity and connectivity. The value of the world resonates in the syncopations and discordant harmonies that entangle and divide the planetary community—the orchestral community of evolution. Insofar as this community is never attained, the discordant harmony of its resonance must be hoped for in the striving and struggling of practice, the straining of training. The practice of listening is a practice of hope. "Music is the art of the hope for resonance: a sense that does not make sense except because of its resounding in itself."[85]

There is no end to practice, no culmination or summit that justifies terminating your training. If there is a final stage in training, it is not when you stop practicing, but when you start practicing like yourself, not like the authorities from which you took up your training regime. This can take a long time. As the jazz trumpeter Miles Davis is reported to have said, "You have to play a long time to sound like yourself."[86] At some indeterminate point, you stop sounding like your teacher. You become immune to the influenza of influences. Authority, sovereignty, and power dissolve, and that is when you begin moving on. That is when you become possible, when you start looking good and sounding good. This resonates with Catherine Malabou's comments about the final stage of training as a woman philosopher, training to affirm the feminine in sexual and ontological difference.

> There comes a time to "do without," to leave behind the masculine, the feminine, and all the other models. A time to abandon the question of authority. One only becomes an authority when one decides to mock authority. This is no doubt the last stage in training and perhaps even in life. A time comes when we know that philosophy has nothing more to offer, that it cannot welcome the fugitive essence of women, that gender studies or deconstruction cannot do so either. We must go off our own, move on, break with, clear new spaces, become possible, in other words, give up power. Power can do nothing against the possible.[87]

It is in the last stage of training that training really becomes possible, when you become possible, and perhaps, if we train together, that is when *we* become possible.

NOTES

1. Bruno Latour, "Why Has Critique Run Out of Steam? From Matters of Fact to Matters of Concern," *Critical Inquiry* 30.2 (2004).
2. Peter Sloterdijk, *Nietzsche Apostle*, trans. Steven Corcoran (Los Angeles: Semiotext(e), 2013), 11.
3. Ibid., 12.
4. Sloterdijk, *Critique of Cynical Reason*, 16.
5. Louisa Shea, *The Cynic Enlightenment: Diogenes in the Salon* (Baltimore: Johns Hopkins University Press, 2010), 135.
6. Sloterdijk, *Globes*, 29.
7. Žižek, *In Defense of Lost Causes*, 70.
8. On the popular "coexist" image, see chapter 3 ("After God, After Nature").
9. Haraway, *When Species Meet*, 301.
10. Sloterdijk, *Globes*, 770.
11. Sloterdijk, *You Must Change Your Life*, 110.
12. This pun, coined by Henk Oosterling, is reiterated by Latour with much appreciation for the way Sloterdijk's conception of humans as designers of spheres avoids any human/nature bifurcation. Bruno Latour, "A Cautious Prometheus? A Few Steps Toward a Philosophy of Design with Special Attention to Peter Sloterdijk," in *In Media Res: Peter Sloterdijk's Spherological Poetics of Being*, ed. Willem Schinkel and Liesbeth Noordegraaf-Eelens (Amsterdam: Amsterdam Unversity Press, 2011), 157.
13. Sloterdijk, *You Must Change Your Life*, 3.
14. Ibid., 13, 113–14. On the phenomenology of vertical tensions in religious experiences, specifically those related to mystical practice, see Anthony J. Steinbock, *Phenomenology and Mysticism: The Verticality of Religious Experience* (Bloomington: Indiana University Press, 2007).
15. Sloterdijk, *You Must Change Your Life*, 21. Rainer Maria Rilke, "Archaic Torso of Apollo," trans. Edward Snow, in *The Art of the Sonnet*, ed. Stephen Burt and David Mikics (Cambridge: Harvard University Press, 2010), 230.
16. Sloterdijk, *You Must Change Your Life*, 23.
17. Rilke, "A Walk," 176.
18. Ibid., 633–34. Michel Foucault, "The Thought of the Outside," in *Essential Works of Michel Foucault, 1954–1984, Vol. 2: Aesthetics, Method and Epistemology*, ed. James D. Faubion (New York: New York Press, 1998), 147–69.
19. Sloterdijk, *You Must Change Your Life*, 633.
20. Sloterdijk, *Globes*, 965.
21. Nancy, *The Creation of the World*, 94.
22. Nancy, *Being Singular Plural*, 135, 137.
23. Sloterdijk, *You Must Change Your Life*, 448–49. Sloterdijk mentions that it was Hans Jonas who provided a "forward-looking philosophy for our times" by reshaping "the categorical imperative into an ecological one" (ibid., 448). Describing the ethical demands of the Anthropocene *avant la lettre*, Jonas says that "an object of an entirely new order—no less than the whole biosphere of the planet—has been added to what we must be responsible for because of our power over it." Jonas, *The Imperative of Responsibility*, 7.
24. Sloterdijk, *You Must Change Your Life*, 451.
25. Ibid., 452.
26. Husserl, *The Crisis of European Sciences and Transcendental Phenomenology*, 150.
27. Morton, *Hyperobjects*, 198.

28. Ibid., 171. A charnel ground is a site for the decay and putrefaction of bodies, such as the open-air burial sites ("sky burial" grounds) found among peoples in the Himalayas, where a lack of tillable soil renders underground burial sites scarce. Tantric exercises in Indo-Tibetan traditions include practices for meditating on charnel grounds. "There is something immensely soothing about charnel grounds" (ibid., 126). On charnel grounds in Tibetan Buddhism, see Judith Simmer-Brown, *Dakini's Warm Breath: The Feminine Principle in Tibetan Buddhism* (Boston: Shambhala Publications, 2001), 121–32.
29. Morton, *The Ecological Thought*, 5.
30. Christie, *The Blue Sapphire of the Mind*, 149.
31. Rachel Carson, *Silent Spring* (Boston: Mariner Books, 2002).
32. Aldo Leopold, *A Sand County Almanac and Sketches Here and There* (London: Oxford University Press, 1968), 22.
33. Ibid., 23.
34. Ibid., 34.
35. Ibid., 54, 61–62.
36. Ibid., 126.
37. Ibid., 82.
38. Ibid., 95–100.
39. Ibid., 101.
40. Ibid., 66.
41. Ibid., 96.
42. Ibid., 153.
43. Ibid.
44. Deleuze, *Difference and Repetition*, 52.
45. Leopold, *A Sand County Almanac and Sketches Here and There*, 207. Baird Callicott, always defending Leopold's land ethic, argues that Leopold, although outdated in terms of chaos theory and deconstruction, nonetheless accounts for the kinds of discord and dissonance that they introduce. Callicott, *Beyond the Land Ethic*, 118–37.
46. Ibid., 129.
47. Ibid.
48. Ibid., 149.
49. Ibid.
50. Ibid., 150.
51. Ibid., 155.
52. Jean-Luc Nancy, *Listening*, trans. Charlotte Mandell (New York: Fordham University Press, 2007), 7.
53. Ibid., 6.
54. Ibid., 29.
55. Ibid., 7.
56. Ibid., 8.
57. Nancy, *The Sense of the World*, 8.
58. Ibid., 63. On Nancy's philosophy of touch, see Derrida, *On Touching*.
59. Nancy, *Listening*, 30.
60. Ibid., 1–6, 69.
61. Ibid., 6.
62. Ibid., 32.
63. Ibid., 15.
64. Ibid., 40.
65. Jean-Luc Nancy, *The Muses*, trans. Peggy Kamuf (Stanford: Stanford University Press, 1996), 38.
66. Jean-Luc Nancy, *The Experience of Freedom*, trans. Bridget McDonald (Stanford: Stanford University Press, 1993), 140.
67. Heidegger, "Letter on 'Humanism,'" 269–71.
68. Nancy, *The Experience of Freedom*, 146.

69. Jean-Luc Nancy, *Multiple Arts: The Muses II*, ed. Simon Sparks (Stanford: Stanford University Press, 2006), 17.
70. Nancy, *Listening*, 38.
71. Ibid., 39.
72. Nancy, *Multiple Arts*, 31.
73. Leopold, *A Sand County Almanac and Sketches Here and There*, 203.
74. Nancy, *The Inoperative Community*, 1.
75. Nancy, *Being Singular Plural*, 65.
76. Ibid., 18.
77. Leopold, *A Sand County Almanac and Sketches Here and There*, 126.
78. Nancy, *Listening*, 16.
79. Ibid., 9.
80. Ibid., 27, 31.
81. Ibid., 21.
82. Ibid., 22.
83. Nancy, *Being Singular Plural*, 3. Leopold's beautiful soul syndrome aside, this resonates with Leopold's sense of community, in which it is possible to say "we" or "us" for humans and nonhumans: "By 'us' I mean the birds, the stream, the dog, and myself." Leopold, *A Sand County Almanac and Sketches Here and There*, 63.
84. Nancy, *The Creation of the World*, 42–43, 47.
85. Nancy, *Listening*, 67.
86. Morton, *Realist Magic*, 22, 152.
87. Malabou, *Changing Difference*, 140.

ELEVEN

Indications of an Axial Age

Things changed after Kierkegaard and Nietzsche. That is not to say that Kierkegaard and Nietzsche caused the change. They might have contributed to the change, and no doubt they have caused many changes vicariously through the dissemination of their writings, but they did not act alone. They heralded the change, of which there are numerous causal factors, including but not limited to a few hundred years of renaissance, reformation, and revolution transforming every aspect of society. The change is that human beings began undoing their *oikos*, destroying their home. As Sloterdijk puts it, "the open animal tore down the roof of its old house from the inside."[1] The free, individual human took the reins of being. Having become unbelievably transcendent, the warming mantles of God were gone, as the sun was unchained from a newly unstable Earth, which Copernicus and others sent into a dizzying orbit.

Exposed to the cold outside, Kierkegaard and Nietzsche decided to live—the courage to be in the cold. They decided: to take a leap of faith, to practice self-overcoming. They did not back out of the harsh cold and look for a system that would have everything make sense again, to put everything in its place as if we had not wiped away the horizon. They knew they were shipwrecked, lost and injured, weak and powerless, and somehow that disempowerment was making possible every leap and overcoming. In the growing desert of modernity, of the West, perhaps necessity is the proverbial mother of invention. In the desert of meaninglessness, they found truths that were true for them, and thus disclosed new possibilities for humans to live their truth in all of its absurdity and profound ambiguity, new possibilities for living and dying in the outside. As Nietzsche knew, at some point he too would become like all free-flying birds. He would be "unable to go on and will perch on a mast or a bare cliff-face," but others "*will fly farther*! This insight and faith of ours vies with them in flying up and away; it rises above our heads and above our impotence into the heights and from there surveys the distance," seeing the birds of the future striving further into the unknown.[2] Leaping and flying into new freedoms and intensities, Nietzsche as well as Kierkegaard landed, perched, and indeed, disastrously wrecked, somewhere, it is rumored, among the islands of loneliness.

As Jaspers puts it, "having dared to be shipwrecked," Kierkegaard and Nietzsche became "representative destinies, sacrifices," whose existential experiments opened up possibilities for new ways of being in the world.[3] Opening up new existential possibilities entailed an experience of the relativity of the Western world. Once taken for granted as a guarantor of universal truth, that world now appeared to have its own provincial limits or, at least, it was on the verge of appearing to have limits. A few stalwarts of occidental imperialism and human exceptionalism still refuse to acknowledge these limits. They refuse to acknowledge the emergency of coexistence after the end of the world. To let that emergency emerge requires acknowledging the existence of others outside of the sexist, racist, classist, speciesist, heteronormative, ethnocentric limits of Western thought. It requires the thought to remain exposed to difference, otherness, the outside.[4] Foucault diagnoses this critical condition of Western thought.

> This turning point, on an historical scale, is nothing other than the end of imperialism. The crisis of Western thought is identical to the end of imperialism . . . it is the end of the era of Western philosophy. Thus, if philosophy of the future exists, it must be born outside of Europe or equally born in consequence of meetings and impacts between Europe and non-Europe.[5]

A term for this crisis is "nihilism." As Keiji Nishitani points out, nihilism is "a crisis in the sense that people began to feel a quaking underfoot of the ground that had supported the history of Europe for several thousand years," and furthermore, "it means that life itself is being uprooted and human 'being' itself turns into a question mark."[6] The beginning of any philosophy of the future would thus be a beginning radically open to question, a beginning that no longer takes for granted the foundations established at the beginning of Western civilization with the emergence and subsequent convergence of biblical religion and Greek philosophy. It would be a philosophy that practices ecognostic attunement to the strangeness and uncertainty of connections between the myriad beings of planetary coexistence.[7] Foucault's comment can be ecologically supplemented to include nonhumans: If there is a philosophy of the future, it must be born in consequence of meetings and impacts not only between Europe and non-Europe but also between humans and nonhumans.

While it was becoming apparent to existentialists that God, nature, humans, and the world were coming to an end, it was also becoming apparent that those foundational realities were not so foundational. Those origins were not so original. They began in the middle of things, conditioned by countless factors, including non-European and nonhuman elements all along. Gradually, in ancient societies, the spheres of God and nature, humans and world, occidental and oriental, were built up while being retroactively designated as given, natural, normal, or original. Their beginning constitutes what Jaspers calls the "Axial Period" or "Axial Age" (*Achsenzeit*), a period of time around the sixth century BCE, a turning point or "axis" (*Achse*) with somewhat fuzzy boundaries, running approximately from 800 to 200 BCE.[8] With the breakdown of traditional and modern immune systems under conditions of ecological emergency, it is not unreasonable to consider the possibility that humanity is somewhere near—before, on, or past—the threshold of a new beginning, a new axial age. I am not talking about the new age of the New Age movement, a utopianism that is indicative of the transcendent "post-conflict" aspirations of much contemporary spirituality, which is making us stupid, selfish, and unhappy.[9] I am talking about a beginning that gets in the political fray and affirms coexistence, as in Hannah Arendt's concept of natality, which indicates that the possibility of a new beginning for all humanity—a "paradoxical plurality of unique beings"—is given with the birth of each human being, as each is singularly unique and irreplaceable, irreducibly human yet entirely different from all other humans.[10] This "new beginning inherent in birth can make itself felt in the world only because the newcomer possesses the capacity of beginning something anew, that is, of acting."[11]

Opening the possibility of acting and being acted on differently, natality bears the possibility of new beginning. With the breakdown of the categories built up during the axial period, is there any indication that, perhaps, a second axial age is beginning, a coexistentialist axial age or, more simply, co-axial age? Although there might be a metaphorical relationship with coaxial cable that is instructive in some way, a coexistential axial age is not to come. The possibility of a new historical axis is rendered senseless by ecological emergency. If some new age is arriving, it provides no axis. It would be an axial age without axes (*Achsenzeit ohne Achsen*), without foundation, ground, horizon, or turning point. It is an age of wildly asymmetrical relationships between humans, approaching a population of almost 7.5 billion, and nonhumans, trillions of trillions of beings whose countlessly different ages put an unbearable weight on any attempt to demarcate a new age, new axis, turning point, or posthistorical or postnihilistic era.

AXIAL EXISTENTIALISM AND NIHILISM

Existentialism and the axial age go hand in hand. To account for one without the other is to fail to understand both. It is no mere coincidence that the person who coined the term *Existenzphilosophie* also coined the term *Achsenzeit*. Similarly, Hans Jonas found it to be no mere coincidence that his understanding of Heidegger's existential phenomenology facilitated his understanding of the continuity of thought from the ancient world to the technologically enframed world of the twentieth century. In particular, Jonas saw this in his study of Gnosticism—an ancient nihilism, for which the body (*soma*) is a tomb (*sema*) in an evil world.[12] Gnosticism for Jonas parallels modern nihilism, for which the world is not evil but completely meaningless, such that there is no world, only meaninglessness, and not even that. Existentialism provided the key to researching this connection.

> What had happened was that Existentialism, which had provided the means of a historical analysis, became itself involved in the results of it. The fitness of its categories to the particular matter was something to ponder about. They fitted as if made to measure: were they, perhaps, made to measure?[13]

In short, the answer is yes, they were made to measure. Gnosticism, modern nihilism, and existentialism all share nihilistic tendencies.

Existentialism is so effective for analyzing the history of thought from classical to modern traditions because existentialism is complicit in the same trend toward nothing. This is not to say that nihilism exhaustively characterizes Gnosticism or existentialism, but nihilism is part of their shared legacy. Moreover, one can extrapolate from the account that Jonas gives and consider the possibility that it is not only Gnosticism but all philosophical and religious traditions of the axial age harbor nihilistic tendencies, which is to say, otherworldly, life-negating tendencies, which paper over the radical alterity of things (their emptiness or *meontic* nothingness) and thus render it null and void (nihility, *oukontic* nothingness).[14] For example, the emancipatory and salvific aims of *moksha* and *nirvāna* in Hindu, Jain, and Buddhist traditions are open to otherworldly interpretations, as are the transcendent God of biblical religions and the transcendent Good of Platonists. Even traditions intently focused on immanence have nihilistic tendencies, for instance, Daoists retreating into quietism or Confucianists effacing the myriad things by assimilating them into rigid hierarchies. The point here is not to criticize axial traditions for their nihilism, as if it is possible or desirable to disentangle oneself from any and all of these traditions and judge them from the outside. All religious traditions have "problems and promise" when it comes to understanding and affirming ecological coexistence.[15] The point is that there is a strange continuity between existentialism and the traditions that emerged in the axial period.

The existentialist movement seems to mark the culmination—or, perhaps more accurately, the shipwreck—of a way of being that characterizes the axial age, beginning almost three thousand years ago with the inception of philosophy and science (Thales, Pythagoras, Plato) and the so-called world religions, i.e., the literate religions (Judaism, Buddhism, Hinduism, Jainism, Confucianism, and Daosim, with Christianity and Islam appearing later, in the first and seventh centuries CE, respectively). World religions stand in contrast to the place-based oral traditions of indigenous peoples, which situate the roots of religion much earlier than axial age religions, arguably extending throughout all of human evolution.[16] The axial age marks the beginning of a way of being that affirms the existence of human beings as such, that is, as instances of a universally applicable category of humanity, exemplified in "paradigmatic individuals" like the Buddha, Confucius, and Socrates, with Jesus following shortly after the axial period.[17] In some sense, those paradigmatic individuals are existentialists *avant la lettre*, axial existentialists. Their similarity to modern existentialists is also their difference, as they appear on different edges of the same history, birthing the human condition in the world during the axial period, and mourning it in the wake of modernity.

The axial period is constitutive of universal humanity, which is itself grounded in other universal categories, like nature, God, and world (hence the appellation, "world" religions). This universalism is persistently haunted by nihilism, the same nihilism haunting coexistence today, as Jonas noted. Universal ideals—human, nature, world, God—threaten to efface the specificity of actually existing beings, assimilating things into supersized containers, grinding things into grounds and backgrounds, and reducing the body to an escape pod to reach otherworldly fulfillment. The axial age was facing an emergency of worlding not altogether unlike that of today, attempting to construct a global immune system, attempting to build a world amid proliferating differences. In the axial age, those differences were primarily philosophical, cultural, and religious differences oriented around ancient imperial and agricultural modes of existence. In the Anthropocene, even more differences are arriving on the scene, differences traversing species, sex, race, gender, class, ethnicity, ability, geography, and more, including the kind of philosophical, cultural, and religious differences that developed in the axial age, and all of these various differences are today oriented around multicultural, postindustrial modes of existence.

The events of the axial period exhibit a "mystery of simultaneity," as the trend toward universality unfolded in relatively independent invention across the Mediterranean, the Middle East, India, and China.[18] Egypt could also be included, although it is perhaps more "proto-axial" compared to the universality of subsequent classical traditions.[19] This mysterious simultaneity can remain an open question for now, and indeed so can the demarcation of an axial period at all. For Jaspers, this period is an empirical fact, but facts remain open, exposed to multiple interpretations. "Every single fact," says Jaspers, "remains inexhaustible and subject to further interpretation," which means "in the end it is always the *unattainable limit* of methodical research."[20] A fact has compelling possibility, pushing inquiry to its limits. While this might seem obvious, it is worth emphasizing, lest theories and facts be taken too seriously, mistaking the menu for the meal, or mistaking forks and knives for a meal or, even more strangely, mistaking the meal (as it appears) for the meal (in its irreducible specificity)—in the Heideggerian terms of object-oriented ontology, mistaking *vorhanden* (present-at-hand) connections for *zuhanden* (ready-to-hand) realities.

"Theorising," as Jaspers puts it, "continually leads us astray. We tend to take the conceivable (the possible) for reality; to mistake what is unverifiable (and random) for what is verifiable (and can be determined); and finally to take what is simply construed as basic and presented metaphorically, for the actual objective realities."[21] The axial age should not be mistaken for a rigid foundation, an absolute point of reference. It is a fact, but not a fact with power. It is a fact with possibility, open for diverse interpretation.[22] The question is how this possibility is taken up in our training, how the truth of the axial age becomes an existential truth, a truth that is true for us. As Jaspers says, it is a question of how the axial age articulation of universality "becomes a concrete reality for us," and "us" in this context includes Jaspers and anyone else whose existence is situated after the audacious shipwrecks of Kierkegaard and Nietzsche.[23]

The axial age articulation of universality was a "transcendental breakthrough," which radicalized previous modes of interpretation found in archaic societies, interpretive methods that can be described as "myth with an element of reflective theory in it," in short, "mythospeculation."[24] The radical mythospeculation occurring in the nascence of philosophy, science, and world religions opened a capacity for contemplating the whole, observing the perfectly round sphere of being, and with that cosmic-ontological perspective, the axial age involved the establishment of the universal categories that have formed, reformed, and deformed human existence since. "In this age were born the fundamental categories within which we still think today, and the beginnings of the world religions, by which human beings still live, were created. The step into universality was taken in every sense."[25]

The step into the universality was the first step into globalization, what Sloterdijk calls the "cosmic-uranian" phase of globalization: universalization as globalization.[26] It is a step into the sphere of universal truth and a step out of the particularity of local, place-based myths. This is an ongoing process, by which people voluntarily or through force abandon the immune systems of local spheres for the promise of greater immunity in the encompassing embrace of universality. Stepping into universality requires pulling yourself away from your local community. The more you pull away, the more they appear like the "they," a herd, the unwashed masses. Moreover, this also requires pulling yourself away from yourself, your own ignorance, sinfulness, or undisciplined habits, and it requires pulling yourself away from the suffering of the world through some sort of redemptive event, whether this-worldly or otherworldly redemption.

The universality of axial age spheres was never merely metaphysical or theoretical. These sphere constructions also involve material conditions, as they took place in agricultural societies existing in and among empires. Those material conditions put very real limitations on the ideals of axial age universality. It is difficult to build heaven on Earth, indeed, impossible. While the attempt to do so might be inevitable or even beneficial, the fantasy of achieving it is delusional, and an attachment to such finality leads to intolerance toward difference and otherness. At worst, this is a matter of violent extremism, and at best, it is a symptom of beautiful soul syndrome, which retreats into a false sense of interiority ("in here") so as to avoid the messy world ("out there"). Such attachment to final truth must be abandoned. "Nowhere on earth can we find final truth, authentic salvation," and any attempt to find and achieve such pure ideals is doomed to fail: "The Axial Period too ended in failure. History went on."[27] History went on, indeed.

The events of the axial age gave rise to history: the birth of history out of the mythospeculative spirit of cosmic globalization. The radical mythospeculation of the axial age led to its own undoing. In terms of Sloterdijk's analysis, the macrospheric intimacy of the axial age led to its own failure, because contemplation of the perfect sphere of being meant simultaneously being outside of the sphere. The cost of radical mythospeculation is the dissociation of humans from their existential closeness, the "betrayal of humanity's existential place": "What Heidegger would later term forgetfulness of being [*Seinsvergessenheit*] already begins with the ancient instructions for viewing the orb blissfully from without," instructions for "lordly observation of the outside that would one day lead to polytechnical dreams of control and the tyranny of knowledge over concretely interpreted life as a whole."[28] Throughout the processes of globalization since the axial period, history has been a matter of competition between universalisms, that is, a competition between globes, leading to the last hundred years of world war, cold war, and global war on terror. "From one point of view," Jaspers observes, "the disintegration and reestablishment of great empires has constituted history ever since the end of the Axial Period."[29] Over the last five hundred years, as the failure of cosmic globalization gave way to terrestrial and electronic attempts at macrospheric construction, from colonialism and industrialization to telecommunications and social media, this historical process has assimilated the planet into the world-interior (*Weltinnenraum*) of capital. The success of capitalism over communism after the cold war marks the end of history, as Francis Fukuyama (in)famously declared, announcing that the "good news has come" as liberal democracy and free market economics have gone global.[30]

Terrestrial and electronic globalization tore down the universals of the axial age to build a climate of overconsumption, a global greenhouse which makes room for tiny islands of pampered pleasure in a vast ocean of injustice and devastation. Existentialists bore witness to the shipwreck of nature, God, and world. They did not witness with critical distance, judging and observing as if from outside. They became shipwrecked themselves along with nature, God, and world. They bemoaned the construction of the global hothouse, a Crystal Palace, recognizing it as another heaven-on-Earth fantasy denying finitude, uncertainty, death, and the unbearable reality of things.[31] They were attempting to let the emergency of coexistence emerge, letting the suffering hurt, practicing to become attentive and sensitive again in the remains of the homogenizing hegemony of "world." They were waking up from the nihilistic fantasies of globalization, which Irigaray accurately depicts as an "anaesthetic world," in which "our greatest pains or greatest joys are abolished by calculation," an annihilation that "kills insensibly. Irremediably. The more you go on producing fantasies, your preferred anaesthetics, the greater the danger: avoiding the passage through suffering which could still save."[32] The passage surely ends in shipwreck, but it is only in the wreckage that there is any hope for renewal. It is only in the emergency of coexistence that anything can emerge.

To turn away from the horrifying suffering and wreckage of contemporary coexistence is to perpetuate the same nihilistic tendencies that have been colonizing the planet for millennia. This is not only the mode of being performed by the corporate executives frequently demonized by environmentalists. It is also the mode of being performed by environmentalists themselves, when they are optimistic about "solving" problems and "fixing" the climate, or when they seek "sustainability" in the Anthropocene. What is called for is quite different. It looks less like learning sustainable design in the Anthropocene and more like "learning to die in the Anthropocene."[33] It looks like the emergence of an emergency that calls humans to be human—again and always only for the first time—to become ecological, feminist, queer, prismatic, psychedelic hyposubjects. Shipwrecked, the universals that have traversed the planet from the axial age to existentialism are interrupted, broken, and scattered into an array of differences—the remains of planetary coexistence. While the construction of the Crystal Palace continues, unabated and even quickened by its certain destruction, the scattered remains are piling up, heaps upon heaps of humans and nonhumans, swarming, assembling intimate alterities, worlding after the end of the world.

AFTER HISTORY

Just like small mammals survived the K-T mass extinction event 65 million years ago, continuing the evolutionary lineage that led to humans, who or what among us will continue to exist in and as the swarm of coexistence in the wake of the current extinction event? Will humans live on? What heap(s) of planetary remains is or are coming to is an open question. It is a coexistential question. If existentialism facilitated an understanding of the first axial age by bearing its wreckage, perhaps coexistentialism provides indications of the arrival of a new axis, a second axial age in which humans take up responsibility for bearing the unbearable, for carrying the weight of every other. After history, after the end of the word, are we arriving at a new turning point? The answer is no, although this "no" means different things in different contexts. There are at last two different ways to say "no" to a second axial age, one is an invalid reason coming from cynical criticism and the other a much more compelling consideration coming from deconstruction via Derrida and Nancy. It matters which "no" says "no."

According to the cynical attitude currently reigning in academia and politically correct discourse, we cannot say "we." In that case, we are definitely not arriving at a new turning point, because "we" is another tool of the hegemonic homogeneity of globalization. The persistence of "we" is the persistence of the same old problems. It is not the arrival of a new age. From this perspective, we should be more reflexive instead of presupposing some collective "we" that is supposed to be obvious or naturally given. Any strategic solidarity provided by subjective plural pronouns is viewed more like totalitarian solidity, which suppresses the hybridity and fluid dynamics of identity formation. In other words, contemporary cynicism assumes that "we" only ever means one thing. It means everyone is the same, and furthermore, it is assumed that sameness is always a simple identitarian equivalence that can be presented or represented to consciousness. Both assumptions are wrong. The word "we" does not necessarily imply sameness, and "sameness" does not necessarily imply an essentialist identity. If it implies essence, it is a weird essentialism for which the continuity of some kind of thing (humans, bicycles, ferns, solar systems, animals, stars, etc.) is an irremediable strangeness that is compellingly close yet withdrawn from all access.[34]

This cynical attitude is more ironic than it admits. In developing a knee-jerk reaction that criticizes use of the word "we" for perpetuating the violent tendencies of globalization, the contemporary cynic is exhibiting the same self-defeating tendency that they pretend to combat, the self-defeating tendency of radical mythospeculation that initiated the cosmic phase of globalization. The reflexivity whereby the perfection and universality of being became disclosed was also the reflexivity whereby humans betrayed their existential place and began judging it as if from outside. That is the position of the cynical critic today, retreating away from the world so as to judge those in it for their failures and successes in meeting the impossible ideals of perfection. Axial traditions emphasized ideals of universality, whereas contemporary theory and political correctness emphasize ideals of specificity and particularity, but those differences matter little compared with the underlying attitude of beautiful soul syndrome: betraying your existential place by fleeing into a false sense of interiority to seek your perfect ideal, and judging and chastising those who fail to meet your impossible demands.

The fact that cynics are admonishing people for using terms of alliance and solidarity (e.g., "we") is a good indication that we are not at a turning point, not because there is something inherently exploitative about saying "we," but because of the cynical attitude that betrays existential closeness and maintains the reflexive distance that opened up during the axial age. If we can no longer say "we," then perhaps we can embrace that hypocrisy of being us and not us at the same time, becoming a community without community or, as Alphonso Lingis says, a "community of those who have nothing in common."[35] Seeking moral purity, the cynic cannot abide that gap. To mind that gap would be to get involved with the ambiguity and uncertainty of coexistence, where nothing is as it seems and everything is strange. To fail to mind that gap is to abandon any possibility for coexistence, retreating instead into a war of foams where every sphere seeks recognition for itself, where every eye rolls at any attempt to build cooperative asceticisms across planetary differences. For instance, consider the cynicism with which some people express greetings for New Year's celebrations.

In the United States, a conventional way to wish someone a festive holiday celebration of the New Year is with this phrase, "Happy New Year." However, the calendrical system for which the first day of the month of January marks the arrival of a new year is only one calendrical system among many, and over the last several centuries, that calendrical system (the Gregorian calendar of Christendom) has been imposed upon all of the world's people through cosmic, terrestrial, and electronic phases of globalization. If you do not want to celebrate a global imposition of a Christian calendar on the rest of the planet, then you should not say a sincere "Happy New Year." Such sincerity indicates your complicity in a violent system. It contaminates your moral purity. To stay pure, you should add a cynical tone to it to indicate that it is not really *the* New Year but just one way to imagine the temporal cycles of terrestrial life. You should say something like, "Happy Gregorian New Year" or "I hope you're enjoying the beginning of another Gregorian cycle." You can be even more cynical, "Happy arbitrarily selected holiday that perpetuates violent power dynamics," "Happy New Year if you are stupid enough to believe in that," or if you decide to simply not utter any New Year's greetings, you can respond to someone else's greetings by saying, "It's not *my* New Year" and/or just roll your eyes.

You might bemoan the globalization of Christendom and its calendar, but if you are saying "Happy New Year" to someone for whom the Gregorian calendar marks the beginning of another year, adding the phrase "Gregorian" to your greeting is not thereby subverting globally imposed temporalities. It is just a way of retreating into your beautiful soul. What such a performance does is situate your subjectivity in a realm of those who know better, those who have looked underneath or beyond the veneer of holiday celebration and found oppressive powers and repressed histories, which are not hard to find. You are not wishing me a happy holiday at all. You are merely expressing your refusal to admit your participation in anything that sullies your beautiful soul, your refusal to admit your complicity in the messy mesh of coexistence. If it does not meet your ideal of moral purity, then you will be derisive and dismissive toward it. Rather than subverting global hegemony, this attitude further entrenches it, still taking away the truth of other calendars (e.g., Chinese, Indian, Aztec, Muslim, etc.). If the globalized Gregorian calendar is supposed to be *the* universally true calendar and each other calendar considered *a* calendar—relative, regionally valid—contemporary cynicism does not refute the demotion of regional calendars to merely relative status. Cynics just want the universal calendar demoted as well, so that no calendar gets to be *the* calendar.

There is perhaps no better way to affirm your allegiance to the Gregorian calendar than by treating calendars as relative assessments of temporal cycles. That is precisely the point of the Gregorian calendar. Calendars are relative, and thus they can be synthesized into meta-calendars in the way that the Gregorian calendar integrates solar and lunar frameworks. If you really wanted to counter the global imposition of Christendom, you would admit that the Gregorian New Year is in fact *the* New Year, and that does not exclude other New Years from also being *the* only New Year. If you do not accept the Gregorian relativizing of calendars, then you must affirm that each calendar is *the* calendar, each time uniquely and irreplaceably. Similarly, if you reject the globally imposed "we," which homogenizes or excludes the differences of individuals and of every particular group or local "we," then you must not simply critique uses of "we." That critical gesture joins the global hegemony in relativizing local groups, leaving humans as lonely individuals by preventing each and all from becoming *the* collective. A more effective way to reject the globally imposed "we" is to affirm that every "we" is in fact *the* "we."

"We" is not just a hegemonic social construct effacing difference. It is us, whoever we are. Saying "we," as Nancy reminds us, "expresses 'our' being divided and entangled."[36] We can "*say we for all being, that is, for every being, for all beings one by one, each time in the singular of their essential plural.*"[37] We cannot completely escape "we," and we cannot escape us, however divided and entangled, singular and plural. The cynical attitude for which we cannot say "we" is thus another untenable nihilism that attempts to escape from the swarm of coexistence. We can and must say "we," each time hypocritically, as we are never simply us, and no single *one* of us is ever merely one of *us*. After the end of the world, cynical criticism of those who judge the messy world from their own private purity is giving way to sincere hypocrisy, a way of being that is inescapably inside the paradoxes of coexistence. Does this mean, then, that we are at a turning point, entering a new era, a second axial age, in which the position of the contemplative observer—which first opened with axial age mythospeculation and has its remains in the cynicism of contemporary relativism and nihilism—is giving way to the sincere hypocrisy of attunement with strange strangers? No. We are not at such a turning point, but it is not because we cannot say "we."

We can and must say "we," but we are not at a turning point. We are in the Anthropocene. Whether you consider it a boundary event or an epoch, the Anthropocene marks a gradual shift of geological epochs, which is happening at a far larger scale than what is meant by a historical turning point or cultural axis. Indeed, the scale of that geological transformation weakens any attempt of humans to tell their own story and define themselves on their own terms. Nonetheless, if Jaspers and other existentialists were capable of discerning the birth of the axial age insofar as they were living in the global nihilistic breakdown that was initiated in that age, it is not unreasonable to consider the hypothesis that the transition to a more ecologically oriented existentialism—coexistentialism—provides an indication of a new age, a postnihilistic or posthistorical period in which the theoretical reflexivity of mythospeculation gives way to the sincere hypocrisy of ecological attunement. Although that hypothesis probably warrants rejection, it is important to consider, as it bears upon the importance of remaining open to the question of what remains to come. What is this world coming to, this world which is not one, this world without world? "*Dans quel(s) monde(s) vivons-nous?* [*What's this (these) world(s) coming to?*]"[38]

Going postal is not the answer. That would amount to more of the same autobiography of human existence that began in the axial age, the same story about human, God, world, and nature weaving through so many reiterations as it spread around the planet during the last three thousand years. While cynical critique goes postal on this autobiography and thus continues its saga, the work of deconstruction welcomes something different. Consider Derrida's remarks on a turning point.

> I would therefore hesitate just as much to say that we are living through a historical turning point. The figure of the turning point implies a rupture or an instantaneous mutation whose model or figure remains, precisely, to be questioned. As for history, historicity, even

> historicality, those motifs belonging precisely—as we shall see in detail—to *this* auto-definition, *this* auto-apprehension, *this* auto-situation of man or of the human *Dasein* as regards what is living and animal life; they belong to this auto-biography of man, which I wish to call into question today.[39]

The turning point of which Derrida is speaking is exactly what is at stake in the transition to an affirmation of coexistence, an alternation to a more extended, open, and hospitable sense of being-with, in other words, a deconstruction of being-with, which is a practice of deconstruction as radically open being-with. In particular, Derrida is talking about an alternation in "the being-with shared by the human and by what the human calls the animal, . . . what we are still daring, provisionally, to name in general but in the singular, the *animal*."[40]

This transition to a more extended and shared being-with is of the utmost importance for humans, animals, and all of planetary coexistence, as the boundary between the human and animal—or human animals and nonhuman animals—is not alone. That boundary intersects with but is not parallel to boundaries between culture and nature, humans and life, animate and inanimate, man and women, transcendence and immanence, intelligibility and sensibility, and indeed, that boundary is entangled with all of the basic categories inherited from the axial age. The opening of the "co-" in coexistence is more urgent and distressing than the limited human concerns at stake in historical turning points. It is nothing other than ecological emergency. The intersection of the human scales with various nonhuman scales of existence renders this alternation of being-with unrecognizable as a historical turning point. It interrupts the autobiography of universal humanity with the strange stories of a universe of intimate alterities.

After the end of the world, humans and the axial age traditions that carried them are powerless, rendered weak, lame, and hypocritical by the realities of the Anthropocene. Vastly outnumbered, humans can give up. Human *Dasein* can relax. You can let the ecological swarm carry you away, which means getting carried away and rightly fascinated and terrified at who we are. By saying "relax" I mean give up power, becoming hyposubjective, which does not necessarily imply remaining calm. I mean abandoning the fantasy of power over others, intensifying and multiplying ways of thinking, feeling, and acting that arise in vulnerable exposure to others. Ecological attunement ruins the project of self-naming, but not because it warrants a prohibition against self-naming, the way a critic would dismiss a self-naming project by pointing out the inequalities and exclusions inherent in it. Ecological attunement ruins the project of self-naming by introducing an exorbitance of other expressive entities, each with its own timeline.

If we are entering a new historical period, it is only in the sense of entering a period in which we do not have the time to declare a new historical period, an "age of asymmetry," as Morton describes it, during which humans are outnumbered by the unbearable closeness of crowds upon crowds of nonhumans.[41] In ecological emergency, it is obvious that things do not run according to the rhythms of human temporality. Beings have the time, timing and spacing each other, interrupting humans, who can take these interruptions as opportunities for training, opportunities to attune to the rhythms of things. Attuning to things does not mean interrupting human subjectivity in order to understand the world objectively. Subject-object dualism is one of those axial categories that came to an end. Attuning to things is about resonating, exchanging contacts between intimate alterities. In Irigaray's words, it is about encounters between subjects. "The other asks me to interrupt the composition of my own weaving of time and space, not in order to go from a subjective to an objective perspective—as is the case in a scientific approach—but to be capable of meeting another subjectivity."[42]

There is no longer any stable ground on which to draw a line between historical periods. The ground quakes, shaking and fracturing human timelines: the turning point of a second axial age is interrupted by the end of the 12,000 years of the Holocene epoch, which began after the last glacial period, and that interruption is itself interrupted by a massive transition—the end of the 65 million years of the Cenozoic era, which was the last age of the dinosaurs. There are too many other times to count: reproductive cycles of salmon, cycles of predator-prey relations, migratory cycles, seasons, sunrises, climate cycles, El Niño, and lunar cycles, among others, including the planetary cycles counted in astronomy and counted on in astrology.

If there is any indication of a new axial age, it is located on a timeline under erasure, in a time after time.[43] Humans do not have the time. The irruption of a new age is interrupted by countless temporal schisms—"chronoschisms."[44] Even the time of my cell phone battery's charge gives lie to my attempt to self-define my time, as do the times of the Cenozoic, the Holocene, the orbit of Mercury, the half-life of plutonium-239, the life cycle of a butterfly, election cycles, and the cycles of multiple calendars. Countless loops of prehistories, histories, posthistories haunt any announcement of posthistory in general or a posthistorical era in the singular. If there is a new axial age, it is an asymmetrical age in which humans are overwhelmed with the differences of 1 + *n* entities, the differences of at least two worlds. Irigaray thus considers that a "third era of the West," following the classical and modern era, might be "the era of the *couple*," the era of 1 + *n* beings, which would herald the feast of sexual difference, "the festival of the world," this world which is not one.[45] The advent of the festival of difference is a turning point only in the sense that it irrevocably interrupts any demarcation of a turning point. In other words, it is a turning point without a turning point. To rephrase Jaspers, it is an *Achsenzeit ohne Achse*, an axial age without axes. History went on after the axial age, and something will go on again after the end of the world. Regardless of what these worlds are coming to, you and I are called on: "you must go on, I can't go on, I'll go on."[46]

We are going on without axes. The axles on the automobile of the historical autobiography of humans broke under the weight of the proliferating hyperobjects of the Anthropocene. This provides an unprecedented opportunity for humans to abandon their nihilistic vehicles and start companioning with others, caring for intimate alterities. However, precisely insofar as this turning point is without axes (grounds, foundations, horizons, nature, God, etc.), this festival is not a clearly posted achievement, neither postnihilist nor posthistorical. We are not free of the nihilistic impulses that have been creeping around the planet throughout cosmic, terrestrial, and electronic histories of globalization. The Crystal Palace continues the nihilistic project, as does the beautiful soul syndrome of critics who judge the palace from the outside instead of giving up power and getting involved. We can continue on this nihilistic path toward extinction, or we can affirm coexistence. This is the concluding message that Clayton Crockett and Jeffrey Robbins express in their proposal for a new materialism. They remind us that, "if we want our civilization to live on a little longer we will have to recognize our coexistence with and in Earth, which is not simply the ground of our being, but the dynamic intensity of our life and thought."[47] In other words, if we want to persist in our dwelling on Earth, we have to get in the fray, carrying and getting carried away by the myriad beings in the swarm of coexistence.

NOTES

1. Sloterdijk, *Bubbles*, 23.
2. Nietzsche, *Daybreak*, 228–29 (#575).
3. Jaspers, *Reason and Existenz*, 38.
4. Foucault, "The Thought of the Outside."
5. Michel Foucault, "Michel Foucault and Zen: A Stay in a Zen Temple," trans. Richard Townsend, in Michel Foucault, *Religion and Culture*, ed. Jeremy R. Carrette (New York: Routledge, 1999), 113.
6. Keiji Nishitani, *The Self-Overcoming of Nihilism*, trans. Graham Parkes with Setsuko Aihara (Albany: SUNY Press, 1990), 173.
7. On ecognosis in contrast to agnosticism, see chapter 3 ("After God, After Nature"). I adapt the term "ecognosis" from Morton, *Dark Ecology*, 5.
8. Karl Jaspers, *The Origin and Goal of History*, trans. Michael Bullock (London: Routledge, 2010), 1.
9. Webster, *Dispirited*, 21. For more on Webster's account of how contemporary spirituality is making us "stupid, selfish, and unhappy," see chapter 3 above ("After God, After Nature").
10. Hannah Arendt, *The Human Condition* (Chicago: University of Chicago Press, 1958), 176.

11. Ibid., 9. This sense of natality and beginning resonates with Derrida's claim about mortality and ending. "For each time, and each time singularly, each time irreplaceably, each time infinitely, death is nothing less than an end of *the* world." Derrida, "Rams," 140. I elaborate on this end of the world in chapter 9 ("Becoming Worldly").

12. Jonas, *The Phenomenon of Life*, 13.

13. Ibid., 212.

14. On the distinction between nihility and emptiness (*śūnyatā*), see Nihitani, *Religion and Nothingness*, 77. On the distinction between oukontic and meontic nothingness, see Tillich, *Systematic Theology: Volume 1*, 188. I discuss these terms in chapter 9 above.

15. This phrase comes from John Grim and Mary Evelyn Tucker, two founders and leaders of the field of religion and ecology. John Grim and Mary Evelyn Tucker, *Ecology and Religion* (Washington, DC: Island Press, 2014), 13.

16. This is Robert Bellah's argument. Religion (not the world religions in particular) began in the Paleolithic period with the arrival of *Homo sapiens*. Robert Bellah, "I still believe that religion in any intelligible sense does not precede our own species and so arose in the Paleolithic period." Robert N. Bellah, *Religion in Human Evolution: From the Paleolithic to the Axial Age* (Cambridge: Belknap Press, 2011), 709–710n2.

17. Karl Jaspers, *Socrates, Buddha, Confucius, Jesus: The Paradigmatic Individuals*, trans. Ralph Manheim (San Diego: Harcourt Brace & Company, 1962).

18. Jaspers, *The Origin and Goal of History*, 13.

19. Bellah, *Religion in Human Evolution*, 243.

20. Karl Jaspers, *Philosophy of Existence*, trans. Richard J. Grabay (Philadelphia: Philadelphia University Press, 1971), 67–68.

21. Karl Jaspers, *General Psychopathology, Volume 2*, trans. J. Hoenig and Marian W. Hamilton (Baltimore: Johns Hopkins University Press, 1997), 549.

22. For a variety of perspectives on multiple dimensions of the axial age (philosophical, religious, sociocultural, anthropological) and their relevance for theories of secularization, comparative religious studies, religious violence, and evolutionary models of religious phenomena, see Robert N. Bellah and Hans Joas, eds., *The Axial Age and Its Consequences* (Cambridge: Belknap Press, 2012).

23. Jaspers, *The Origin and Goal of History*, 21.

24. Bellah, *Religion in Human Evolution*, 276.

25. Jaspers, *The Origin and Goal of History*, 2.

26. Sloterdijk, *In the World Interior of Capital*, 9.

27. Jaspers, *The Origin and Goal of History*, 20.

28. Sloterdijk, *Globes*, 79.

29. Jaspers, *The Origin and Goal of History*, 6.

30. Francis Fukuyama, *The End of History and the Last Man* (New York: Avon Books, 1992), xi–xiii.

31. Sloterdijk refers to the global hothouse of the "Crystal Palace" of Dostoevsky in his *Notes from Underground*. Sloterdijk, *In the World Interior of Capital*, 169. Cf. Dostoevsky, *Notes from Underground*, 23–25.

32. Irigaray, *Elemental Passions*, 55.

33. Scranton, *Learning to Die in the Anthropocene*.

34. I discuss such "weird essentialism" (a term from object-oriented ontology) in chapter 8 ("Looking Good"). See also Morton, *Dark Ecology*, 64–65, 70–72.

35. Alphonso Lingis, *The Community of Those Who Have Nothing in Common* (Bloomington: Indiana University Press, 1994).

36. Nancy, *Being Singular Plural*, 65.

37. Ibid., 3.

38. Nancy and Barrau, *What's These Worlds Coming To?*, 7.

39. Jacques Derrida, *The Animal That Therefore I Am*, ed. Marie-Louise Mallet, trans. David Wills (New York: Fordham University Press, 2008), 24.

40. Ibid.

41. Morton, *Hyperobjects*, 159.

42. Irigaray, *Sharing the World*, 86–87.
43. Time after time is what remains of time after the deconstruction of the metaphysics of presence, that is, the deconstruction of the ontotheology of God, nature, world, and human. It is a time of multiple times looping around and interrupting one another in fractal iterations, recursive repetitions of existential facticity. David Wood calls this fractal facticity of multiple temporalities "a co-temporal *fracticity*." David Wood, *Time after Time* (Bloomington: Indiana University Press, 2007), 82.
44. Ursula K. Heise, *Chronoschisms: Time, Narrative, and Postmodernism* (Cambridge: Cambridge University Press, 1997).
45. Irigaray, *An Ethics of Sexual Difference*, 148–49. I discuss Irigaray's conception of sexual difference as feast or festival in chapter 8 ("Looking Good").
46. Beckett, *The Unnamable*, 134.
47. Clayton Crockett and Jeffrey W. Robbins, *Religion, Politics, and the Earth: The New Materialism* (New York: Palgrave Macmillan, 2012), 153.

TWELVE

Coda

Not a typical movement or "ism," and not a clearly definable type, kind, or genre of philosophy and literature, coexistentialism resists being typecast. That is part of its existentialist fashion. When it shows up, it can play many different roles, changing its accents, appearances, and outfits to find different ways of acting and being acted on by others. Yet, it is not simply anything whatsoever, a relativistic category that can be applied haphazardly or in a whim. It is ecological. Coexistentialism is an ecological existentialism, and ecology means far more than a science of organism-environment interactions or a generalized study of interrelationships. "Ecology," as Morton reminds us, "is about intimacy," which is not a cozy familiarity or concordant harmony.[1] Ecological intimacy is more unbearable than that, as it carries and is carried away by swarming singularities—real beings glowing with compelling displays of horrifying and fascinating alterity. "Ecological coexistence is with ghosts, strangers, and specters."[2] Ecological existentialism is about training in being-with. It is about cooperative exercises for attending to strange strangers in the mesh of coexistence, exercises for responding to uncanniness and uncertainty amid radical connectivity. In some ways worlds apart (see chapter 9, "Becoming Worldly"), Morton is not very far from Rose. "Ecological existentialism," in Rose's sense, "responds to the two big shifts in Western thought that define our current moment: the shift into uncertainty and the shift into connectivity."[3]

Ecological existentialism is a renewal of existentialism in an ecological context. It takes up the legacies of existentialism while undoing the lonely human subject of existentialism, redoing subjectivity in a more open and extended sense of being-with that includes all beings, human, nonhuman, and otherwise. Although I speak of ecological existentialism and coexistentialism in the singular throughout this book, it is a singular plurality, if it is at all. Morton's coexistentialism and Rose's ecological existentialism are two among many ecological renewals of existentialist legacies, like Nancy's philosophy of sense, Sloterdijk's spherology, Malabou's new materialism of plasticity, Haraway's companion species cosmopolitics, and more, with no unambiguous boundary that decides who is in and who is out, who fits and who does not. Similarly, there is an open-ended list of predecessors and coexistentialists *avant la lettre,* with phenomenologists of alterity like Levinas and Lingis and feminists of difference like Irigaray and Plumwood standing out, and many existentialist philosophers themselves already expressing ecological intimations. There is much that is shared among those various theorists, and there are also fault lines between them, as indicated in the difference between the accounts of new materialists and object-oriented ontologists regarding the status of beings and relationships (see chapter 9, "Becoming Worldly"). The fault lines multiply, with ecofeminism, queer ecology, eco-phenomenology, ecocriticism, and ecological humanities in the coexistential mesh with object-oriented ontology and new materialism. Those fault lines are a good thing. "It is usually at the edges," as Plumwood observes, "where the great tectonic plates of theory meet and shift that we find the most dramatic developments and upheavals."[4]

The tectonic shift of concern for ecological coexistence under way in recent decades is shaking up the sclerotic categories confining contemporary thinking. The cracks and gaps opening in these tectonic shifts are making room for intimate alterities. Multiple theories and fields of study involved in coexistentialism push one another, in mutual alliance and resistance, sharing the struggle to understand and affirm coexistence during a time of ecological emergency. Affirming emergency does not mean enjoying ecological destruction or being happy about it. It is not about positive thinking. It means that it is better to be in the fray than to be disengaged, living beyond the fray in the private interiority of the beautiful soul or the private property of neoliberal enclosures. It means that you want to be in the swarm of coexistence, bearing its unbearable stress, pressure, and urgency.

> How beautiful to be!
> State of emergency
> Is where I want to be.[5]

Coexistentialism is not going to solve anybody's problems. Far from a panacea, it does not help. It is not an extension of self-help to other-help. It does not make anything okay. I am not okay, and you are not okay either. Coexistentialism does not propose sustainability solutions. If there is hope for solutions, it is a hope that can come only in letting problems be problems, letting the emergency of coexistence emerge. If it cures anything, it cures the delusional fantasy that there is a cure for coexistence.

"Use your head, can't you, use your head, you're on earth, there's no cure for that!"[6]

An invitation and a warning, the word "coexistentialism" is a marker, a key of sorts, describing the overwhelming feeling that you are in the fray, the crowd of coexistence. "We need a term like 'coexistentialism' to describe what it feels like to be a swarming colony: we contain multitudes."[7] Rather than pretending to know what is wrong or what needs to be fixed, you can affirm your uncertain and interconnected participation in multitudes of beings, heaping, piling up, swarming, assembling, stretching out, extending, and also, perhaps, worlding. What if the only problem that needs to be fixed is the impulse to fix problems? It is easy to lose ourselves in fantasies of problem-solving. We do not simply need better problem-solving. Even when avoiding any oversimplification of problems, accounting for all of their unruly, wicked, and super-wicked traits, you might nonetheless fail to attend to your uncanny belonging in and as the swarm, a stranger among other strangers, *partes extra partes*. There is so much talk of super-wicked problems and sustainability solutions, but the feeling of vulnerable exposure in the swarm is much less comfortable to talk about. It is a strange feeling, a little too intimate.

By way of extension outside of the body of this book, in its short "tail" (coda), I will say something a little too intimate, something personal about my way into coexistentialism. I am not intending to render this a personal confession of what is presented impersonally throughout the rest of the book. The point of offering a personal tale in this tail is precisely to indicate that the tail is not outside of the body, but folds back into it like an Ouroboros. I do not stand outside of this book like an observer, communicating about a determinate topic called "coexistentialism." I am intimately involved with it, perhaps pulling people into it that would prefer to abstain from the label. I do not need to say "I" to write about my existence. The authors, writings, ideas, perceptions, affects, relations, and all beings woven into my writing in this book are my personal confession. My intimate disclosure here is that this entire book has been an intimate disclosure, an opening of my vulnerable exposure to others. It is a dissimulated, hypocritical disclosure, to be sure. A more direct disclosure would risk overexposure and indecent exposure.

I hesitate to express anything that seems like a relatively direct disclosure about my personal experience, as such an expression risks reinforcing a false dichotomy between impersonal academic writing and personal experiential writing. Even the driest academic prose is personal, however meager the personality, and even the most personal writing touches on the topics of the highest academic discourses, however vaguely or obliquely. Without perpetuating dichotomies of abstract/concrete or impersonal/personal, I want to adhere to an insistent injunction of existentialism: to live your philosophy, not in the sense of embodying an abstraction but in the sense of practicing philosophy as a matter of life and death. This is livelier than the famous Socratic dictum that the unexamined life is not worth living. If your life is not lively, it is not worth getting philosophical about it. Taking up this existentialist legacy, Alphonso Lingis reminds us that "the unlived life is not worth examining."[8] Writing existential truth entails more than writing about one's own life. It entails writing about a vibrantly lived life, an intense life on the edge, remaining exposed to the strangeness of whatever remains after God, nature, and the end of the world. Remaining exposed, my life is never simply my own, and it is never simply life. My own life is always already entangled with my own death as well as the lives and deaths of the others that I must carry and by which I get carried away. I hesitate to express anything that seems like a relatively direct disclosure about my personal experience, but to follow the injunction of existentialism—practicing philosophy as life and death—a personal tale in this coda must surely include more than a declaration that everything I have said is indirectly personal.

Attempting not to stray far from the ouroboric folding of the tail into itself, my personal tale in this literary "tail" will fold into a literal tail, in particular, the tail of a female mammal, and more specifically, the tail of a domestic short-haired cat with black fur. The chromatic phenomenon of her blackness situates her in "a black or existential ecology," that is, an existential ecology of black bodies, marking her alterity as well as her subordinate status as nonhuman in a speciesist society, which is not incidentally also a racist society, as multiple modes of blackness are suppressed by the white mythology of sovereign presence.[9] This cat has metaphorical implications, but she is indeed a real cat, one that plays, walks, eats, meows, snores, and never uses the litter box when I am nearby. It is not Derrida's cat, although I resonate with much in his account of his relationship with that little cat, and I consider him alongside Levinas and Irigaray as a theorist of coexistentialism.[10] I am not attempting any deconstructive acrobatics here, just a short tale. It is a short tale about a broken tail. It does not seem severe, as she does not exhibit any symptoms that would indicate its severity, such as difficulty walking or urinating, but there is an easily noticeable crook where there was a break in her tail. The break makes her stand out, and what stands out is her opacity, a break in appearance and access, an indication of the irremediable difference of black bodies.

Her time interrupts ours, but we do not know her history. The cause of the break is forever withdrawn. When we (two of us, me and Kim, my partner) met her, it was not clear if she was stray, feral, or a neighbor's cat that spends time outside. Glances from a distance gradually gave way to closer proximity, which provided space for the light touches of tactful contact. Eventually, after weeks of those interactions, as it became apparent that this cat had no home, more specifically, no regular and stable shelter, we invited her inside. Hesitant at first, she very gradually became accustomed to spending time indoors with us throughout the day and began staying throughout the night on a somewhat regular basis. We let her out often, and she kept returning.

You could say that we are in the process of adopting or rescuing this cat, although not without much hesitation regarding ideas of adoption and rescue as they implicate a certain questionable understanding of relationships between humans and animals. Who is adopting who? Who is rescuing who? Who selects who? The usual dynamics of active subjects and passive objects have imploded. She is a strange being for which classical categories like "being" and "animal" seem too fitting to do justice to her or to do justice to our relationship. In short, we are companioning with this stranger, this cat, whose name is not one. For no apparent reason, we have come to speak and write about her with a couple of phonetically indistinguishable names, "misses" and "Mrs.," not to mention an ongoing proliferation of nicknames that would be far too intimate to disclose here.

While we are getting used to each other, habituated and normalized, she nonetheless seems stranger every day, making me question myself, the very existence of organisms, and the unfathomable horror of mass extinction. Furthermore, this strangeness is not without problems. I will restrict myself to a few. First, my apartment complex does not allow pets, giving her an illegitimate status that makes caring for her difficult and risky. Second, she had fleas and continues picking up more around our neighborhood, requiring uncomfortable treatments and bathing, and demanding our obsessive attention to tiny things (fleas, flea dirt, eggs, and larvae) and frequent housecleaning. Third, her love is entirely conditional, dependent upon my actions toward her and, perhaps most of all, dependent on giving her food. We are companions, being together by eating, being with food, being with bread (*cum panis*). Giving up any transcendent fantasies of the unconditioned or unconditional, love is impossibly difficult work, bearing the unbearable needs and demands of the other. The glance of the lover wells up with tears. My heart breaks for this cat, and her illegality and parasites become things I must bear. It is not that she is worth the risk, as if there is some cost-benefit analysis involved, some external frame of reference that can tell me what I should do. There is no calculus of care, no external frame of reference. The world is gone, I must carry her.[11] Why this cat and not another? There is no way of knowing. "What knowledge could ever measure up to the injunction to choose between those whom one loves, whom one must love, whom one can love?"[12] Our companionship is radically contingent. Remaining vulnerable, I do not take her for granted. I take her to heart.

Coexistentialism is about giving up power and becoming possible, becoming hyposubjects, refuting loneliness, and building intimate spaces. Insofar as it is an "ism" resistant to "isms," it is only meant to emphasize that this way of thinking, this style, this practice is never separate from its instantiation amid the exigencies of actual existence. Coexistentialism is not cynical. It is not afraid of getting its hands dirty with "isms" and other labels and abstractions, institutions and histories, powers and possibilities. It is an "ism" without an "ism" insofar as it is an "ism" on the move, an "ism" that gets in the fray, changing hearts, changing differences. An "ism" with movement and changes, coexistentialism is musical. It is about sounding good to and for others, attuning to notes from the underground of coexistence, notes from the subterranean depths of intimate alterities. In the soundtrack of your existence, use terms like "coexistentialism," "ecological existentialism," and "existential ecology," or give them a wide berth. Think with coexistentialist thinkers, or let sleeping dogs lie, dreaming. In any case, we are here, coming and going.

NOTES

1. Morton, *Hyperobjects*, 139.
2. Ibid., 195.
3. Rose, *Wild Dog Dreaming*, 42.
4. Plumwood, *Feminism and the Mastery of Nature*, 1.
5. Björk, "Jóga."
6. Samuel Beckett, *Endgame and Act without Words* (New York: Grove Press, 1994), 53.
7. Morton, "Deconstruction and/as Ecology," 301.
8. Alphonso Lingis, *Abuses* (Berkeley: University of California Press, 1994), 91.
9. An interpretation of the inaccessible and subordinate status of others in terms of "black bodies" is given in Levi Bryant's "existential ecology." Bryant, "Black," 292–93, 301. "Black bodies are not simply entities harboring hidden powers but also entities caught in ecological networks that perpetuate their disadvantage and lack of representation and that are particularly vulnerable to economic, natural, and institutional inequalities and forms of destruction" (ibid., 306). On "white mythology," which entangles the logocentric metaphysics of presence with the white supremacy of Western imperialism and modern colonialism, see Jacques Derrida, "White Mythology: Metaphor in the Text of Philosophy," in Jacques Derrida, *Margins of Philosophy*, trans. Alan Bass (Chicago: University of Chicago Press, 1982).
10. Derrida, *The Animal That Therefore I Am*, 6.
11. This is a paraphrase of the poetic saying of Celan that figures into Derrida's meditations on carrying the other at and as the end of the world. "The world is gone, I must carry you." Derrida, "Rams," 141.
12. Derrida, *Politics of Friendship*, 22.

References

Abram, David. *Becoming Animal: An Earthly Cosmology*. New York: Pantheon Books, 2010.

———. "Between the Body and the Breathing Earth: A Reply to Ted Toadvine." *Environmental Ethics* 27.2 (2005): 171–90.

———. *The Spell of the Sensuous: Perception and Language in a More-than-Human World*. New York: Vintage Books, 1997.

Alaimo, Stacy, and Susan Hekman, eds. *Material Feminisms*. Bloomington: Indiana University Press, 2008.

Allitt, Patrick. *A Climate of Crisis: America in the Age of Environmentalism*. New York: Penguin Press, 2014.

Arendt, Hannah. *The Human Condition*. Chicago: University of Chicago Press, 1958.

———. *The Promise of Politics*. Edited by Jerome Kohn. New York: Random House, 2005.

Aristotle. *Metaphysics*. Translated by Joe Sachs. Santa Fe: Green Lion Press, 1999.

———. *Nicomachean Ethics*. Translated by Joe Sachs. Newburyport: Focus Publishing, 2002.

———. *The Politics*. Translated by Carnes Lord. Chicago: University of Chicago Press, 1984.

Association for the Study of Literature and Environment. "Notes from Underground: The Depths of Arts, Culture and Justice." *ASLE Biennial Conference*. University of Idaho, June 23–27, 2015. www.aslebiennialconference.com/index.html (accessed on February 1, 2016).

Augustine. *The Confessions of Saint Augustine*. Translated by John K. Ryan. New York: Doubleday, 1960.

Bachelard, Gaston. *The Flame of a Candle*. Translated by Joni Caldwell. Dallas: Dallas Institute Publications, 1988.

———. *The Formation of the Scientific Mind*. Translated by Mary McAllester Jones. Manchester: Clinamen Press, 2002.

———. *The Poetics of Space*. Translated by Maria Jolas. Boston: Beacon Press, 1994.

———. *Water and Dreams: An Essay on the Imagination of Matter*. Translated by Edith R. Farrell. Dallas: The Dallas Institute of Humanities and Culture, 1983.

Barad, Karen. *Meeting the Universe Halfway: Quantum Physics and the Entanglement of Matter and Meaning*. Durham: Duke University Press, 2007.

Barrett, William. *Irrational Man: A Study in Existential Philosophy*. New York: Anchor Books, 1990.

Bauman, Whitney A. *Religion and Ecology: Developing a Planetary Ethic*. New York: Columbia University Press, 2014.

Beckett, Samuel. *Endgame and Act without Words*. New York: Grove Press, 1994.

———. *The Unnamable*. London: Faber and Faber, 2012.

Bell, Jeffrey A. "Undoing the Subject: Deleuze and the Makings of a Sustainable Life." In *New Directions in Sustainable Design*, edited by Adrian Parr and Michael Zaretsky, 229–41. New York: Routledge, 2011.

Bellah, Robert N. *Religion in Human Evolution: From the Paleolithic to the Axial Age*. Cambridge: Belknap Press, 2011.

Bellah, Robert N., and Hans Joas, ed. *The Axial Age and Its Consequences*. Cambridge: Belknap Press, 2012.

Belo, Catarina. *Chance and Determinism in Avicenna and Averroes*. Leiden: Brill, 2007.

Bennett, Jane. *Vibrant Matter: A Political Ecology of Things*. Durham: Duke University Press, 2010.
Bergson, Henri. *An Introduction to Metaphysics*. Translated by T. E. Hulme. New York: G. P. Putnam's Sons, 1912.
Björk. *Björk: Archives*. London: Thames and Hudson, 2015.
———. "Jóga." *Homogenic*. London: One Little Indian, 1997.
Blake, William. *The Complete Prose and Poetry of William Blake*. Revised edition. Edited by David V. Erdman. Berkeley: University of California Press, 2008.
Bogost, Ian. *Alien Phenomenology, or What It's Like to Be a Thing*. Minneapolis: University of Minnesota Press, 2012.
Bookchin, Murray. "Social Ecology versus Deep Ecology." In *Environmental Ethics: Readings in Theory and Application*, 5th edition, edited by Louis P. Pojman and Paul Pojman, 241–52. Belmont: Thomson Wadsworth, 2008.
Boyer, Dominic, and Timothy Morton. "Hyposubjects." *Theorizing the Contemporary, Cultural Anthropology*, January 21, 2016. www.culanth.org/fieldsights/798–hyposubjects (accessed on February 3, 2016).
Brody, Robert. "The Existential Genius of Late Woody Allen." *The New Yorker*, July 16, 2015. www.newyorker.com/culture/cultural-comment/the-existential-genius-of-late-woody-allen (accessed on September 3, 2015).
Bryant, Levi. "Black." In *Prismatic Ecology: Ecotheory Beyond Green*, edited by Jeffrey Jerome Cohen, 290–310. Minneapolis: University of Minnesota Press, 2013.
Buber, Martin. *Eclipse of God: Studies in the Relation Between Religion and Philosophy*. Translated by Robert Seltzer. New York: Humanity Books, 1988.
———. *I and Thou*. Translated by Walter Kaufmann. New York: Touchstone, 1970.
Butler, Judith. *Undoing Gender*. New York: Routledge, 2004.
Cain, Clifford Chalmers, ed. *Many Heavens, One Earth: Readings on Religion and the Environment*. Lanham: Lexington Books, 2012.
Callicott, J. Baird. *Beyond the Land Ethic: More Essays in Environmental Philosophy*. Albany: SUNY Press, 1999.
———. *Thinking Like a Planet: The Land Ethic and the Earth Ethic*. New York: Oxford University Press, 2013.
Camus, Albert. *The Myth of Sisyphus and Other Essays*. Translated by Justin O'Brien. New York: Vintage Books, 1991.
———. *The Rebel: An Essay on Man in Revolt*. Translated by Anthony Bower. New York: Vintage Books, 1991.
Caputo, John D. *How to Read Kierkegaard*. New York: W. W. Norton & Company, 2007.
———. *The Insistence of God: A Theology of Perhaps*. Bloomington: Indiana University Press, 2013.
———. *More Radical Hermeneutics: On Not Knowing Who We Are*. Bloomington: Indiana University Press, 2000.
———. *The Prayers and Tears of Jacques Derrida: Religion without Religion*. Bloomington: Indiana University Press, 1997.
———. *The Weakness of God: A Theology of the Event*. Bloomington: Indiana University Press, 2006.
Carson, Rachel. *Silent Spring*. Boston: Mariner Books, 2002.
Casey, Edward S. *The Fate of Place: A Philosophical History*. Berkeley: University of California Press, 1997.
———. "Taking a Glance at the Environment: Preliminary Thoughts on a Promising Topic." In *Eco-Phenomenology: Back to the Earth Itself*, edited by Charles S. Brown and Ted Toadvine, 187–210. Albany: SUNY Press, 2003.
———. *Spirit and Soul: Essays in Philosophical Psychology*. 2nd edition. Putnam: Spring Publications, 2004.
———. *Getting Back Into Place: Toward a Renewed Understanding of the Place-World*. Bloomington: Indiana University Press, 1993.
Cataldi, Suzanne L., and William S. Hamrick, eds. *Merleau-Ponty and Environmental Philosophy: Dwelling on the Landscapes of Thought*. Albany: SUNY Press, 2007.

Christie, Douglas E. *The Blue Sapphire of the Mind: Notes for a Contemplative Ecology*. New York: Oxford University Press, 2013.

Cohen, Jeffrey Jerome. *Stone: An Ecology of the Inhuman*. Minneapolis: University of Minnesota Press, 2015.

Cohen, Jeffrey Jerome, ed. *Prismatic Ecology: Ecotheory Beyond Green*. Minneapolis: University of Minnesota Press, 2013.

Cohen, Jeffrey Jerome, and Lowell Duckert, eds. *Elemental Ecocriticism: Thinking with Earth, Air, Water, and Fire*. Minneapolis: University of Minnesota Press, 2015.

Cohen, Tom, Claire Colebrook, and J. Hillis Miller. *Twilight of the Anthropocene Idols*. London: Open Humanities Press, 2016.

Confucius. *The Analects*. Translated by Raymond Dawson. Oxford: Oxford University Press, 2008.

Coppinger, Raymond, and Lorna Coppinger. *Dogs: A New Understanding of Canine Origin, Behavior and Evolution*. Chicago: University of Chicago Press, 2001.

Crockett, Clayton, and Jeffrey W. Robbins. *Religion, Politics, and the Earth: The New Materialism*. New York: Palgrave Macmillan, 2012.

Curb Your Enthusiasm. "Beloved Aunt." *Curb Your Enthusiasm Episode Scripts*. www.springfieldspringfield.co.uk/view_episode_scripts.php?tv-show=curb-your-enthusiasm&episode=s01e08 (accessed September 3, 2015).

Daigle, Christine, ed. *Existentialist Thinkers and Ethics*. Montreal: McGill-Queen's University Press, 2006.

Darwin, Charles. *The Descent of Man, and Selection in Relation to Sex*. Princeton: Princeton University Press, 1981.

De Beauvoir, Simone. *The Ethics of Ambiguity*. Translated by Bernard Frechtmen. New York: Citadel Press, 2000.

———. *The Second Sex*. Translated by Constance Borde and Sheila Malovany-Chevallier. New York: Random House, 2009.

Deleuze, Gilles. *Difference and Repetition*. Translated by Paul Patton. New York: Columbia University Press, 1994.

Deleuze, Gilles, and Félix Guattari. *A Thousand Plateaus: Capitalism and Schizophrenia*. Translated by Brian Massumi. Minneapolis: University of Minnesota Press, 1987.

Derrida, Jacques. *The Animal That Therefore I Am*. Edited by Marie-Louise Mallet. Translated by David Wills. New York: Fordham University Press, 2008.

———. *The Beast and the Sovereign, Volume II*. Translated by Geoffrey Bennington. Chicago: University of Chicago Press, 2011.

———. *Dissemination*. Translated by Barbara Johnson. Chicago: University of Chicago Press, 1981.

———. *Glas*. Translated by John P. Leavey Jr. and Richard Rand. Lincoln: University of Nebraska Press, 1986.

———. "Interpreting Signatures (Nietzsche/Heidegger): Two Questions." In *Dialogue and Deconstruction: The Gadamer-Derrida Encounter*, edited by Diana P. Michelfelder and Richard E. Palmer, 58–72. Albany: SUNY Press, 1989.

———. *Monolingualism of the Other: Or, the Prosthesis of Origin*. Translated Patrick Mensah. Stanford: Stanford University Press, 1998.

———. *Politics of Friendship*. Translated by George Collins. London: Verso, 1997.

———. "Rams: Uninterrupted Dialogue—Between Two Infinities, the Poem." Translated by Thomas Dutoit and Philippe Romanski. In Jacques Derrida, *Sovereignties in Question: The Poetics of Paul Celan*, edited by Thomas Dutoit and Outi Pasanen, 135–63. New York: Fordham University Press, 2005.

———. *Specters of Marx: The State of the Debt, the Work of Mourning, and the New International*. Translated by Peggy Kamuf. New York: Routledge, 1994.

———. *Spurs: Nietzsche's Styles/Éperons: Les Styles de Nietzsche*. Translated by Barbara Harlow. Chicago: University of Chicago Press, 1979.

———. *On Touching—Jean-Luc Nancy*. Translated by Christine Irizarry. Stanford: Stanford University Press, 2005.

———. "White Mythology: Metaphor in the Text of Philosophy." In Jacques Derrida, *Margins of Philosophy*, translated by Alan Bass, 207–71. Chicago: University of Chicago Press, 1982.
Descartes, René. *Principles of Philosophy*. Translated by Valentine Rodger Miller and Reese P. Miller. Dordrecht: Kluwer Academic Publishers, 1982.
Descola, Philippe. *Beyond Nature and Culture*. Translated by Janet Lloyd. Chicago: University of Chicago Press, 2005.
Dōgen, Eihei. *Moon in a Dewdrop: Writings of Zen Master Dōgen*. Edited by Kazuaki Tanahashi. New York: North Point Press, 1985.
Dostoevsky, Fyodor. *Notes from Underground*. Translated by Richard Pevear and Larissa Volokhonsky. New York: Alfred A. Knopf, 2004.
Doyle, Richard M. *Darwin's Pharmacy: Sex, Plants, and the Evolution of the Noösphere*. Seattle: University of Washington Press, 2011.
Ehrlich, Paul. *The Population Bomb*. Revised edition. Jackson Heights: Rivercity Press, 1975.
Eliade, Mircea. *Images and Symbols: Studies in Religious Symbolism*. Translated by Philip Mairet. Princeton: Princeton University Press, 1991.
———. *The Sacred and the Profane: the Nature of Religion*. Translated by Willard R. Trask. San Diego: Harcourt Brace Jovanovich, 1987.
Fisher, Andy. *Radical Ecopsychology: Psychology in the Service of Life*. Albany: SUNY Press, 2002.
Flohr, Paul Mendes. "Jewish Co-Existentialism: Being with the Other." In *Situating Existentialism: Key Texts in Context*, edited by Jonathan Judaken and Robert Bernasconi, 237–55. New York: Columbia University Press, 2012.
Foucault, Michel. *The History of Sexuality, Volume 1: An Introduction*. Translated by Robert Hurley. New York: Vintage Books, 1990.
———. *The Order of Things: An Archaeology of the Human Sciences*. Translated by Alan Sheridan. London: Routledge, 2004.
———. "Michel Foucault and Zen: A Stay in a Zen Temple." Translated by Richard Townsend. In Michel Foucault, *Religion and Culture*, edited by Jeremy R. Carrette, 110–14. New York: Routledge, 1999.
———. "The Thought of the Outside." In *Essential Works of Michel Foucault, 1954–1984, Vol. 2: Aesthetics, Method and Epistemology*, edited by James D. Faubion, 147–69. New York: New York Press, 1998.
Franklin, Amanda M., Zoe E. Squires, and Devi Stuart-Fox. "The Energetic Cost of Mating in a Promiscuous Cephalopod." *Biology Letters* 8 (2012): 754–56.
Freud, Sigmund. *A General Introduction to Psychoanalysis*. Translated by Joan Riviere. New York: Pocket Books, 1972.
Fukuyama, Francis. *The End of History and the Last Man*. New York: Avon Books, 1992.
Gane Nicholas, and Donna Haraway. "When We Have Never Been Human, What Is To Be Done?" *Theory, Culture and Society* 23 (2006): 135–58.
Gaston, Sean. *The Concept of World from Kant to Derrida*. London: Rowman & Littlefield International, 2013.
Al-Ghazalī, Abū Hāmid. *The Incoherence of the Philosophers*. 2nd edition. Provo: Brigham Young University Press, 2000.
Gibson, James J. *The Ecological Approach to Visual Perception*. Boston: Houghton Mifflin, 1979.
Goldblatt, David. *Art and Ventriloquism: Essays*. New York: Routledge, 2006.
Grim, John, and Mary Evelyn Tucker. *Ecology and Religion*. Washington, DC: Island Press, 2014.
Grosz, Elizabeth. *Becoming Undone: Darwinian Reflections on Life, Politics, and Art*. Durham: Duke University Press, 2011.
———. *Chaos, Territory, Art: Deleuze and the Framing of the Earth*. New York: Columbia University Press, 2008.
———. "Naked." In *Encounters with Alphonso Lingis*, edited by Alexander E. Hooke and Wolfgang W. Fuchs, 119–32. Lanham: Lexington Books, 2003.

———. *Volatile Bodies: Toward a Corporeal Feminism*. Bloomington: Indiana University Press, 1994.

Hadot, Pierre. *Philosophy as a Way of Life: Spiritual Exercises from Socrates to Foucault*. Translated by Michael Chase. Malden: Blackwell Publishing, 1995.

Hall, Matthew. *Plants as Persons: Toward a Philosophical Botany*. Albany: SUNY Press, 2011.

Hamrick, William S., and Jan Van der Veken. *Nature and Logos: A Whiteheadian Key to Merleau-Ponty's Fundamental Thought*. Albany: SUNY Press, 2011.

Hannesson, Rögnvaldur. *Ecofundamentalism: A Critique of Extreme Environmentalism*. Lanham: Lexington Books, 2014.

Haraway, Donna. "Anthropocene, Capitalocene, Plantationocene, Chthulucene: Making Kin." *Environmental Humanities* 6 (2015): 159–65.

———. *The Companion Species Manifesto: Dogs, People, and Significant Otherness*. Chicago: Prickly Paradigm Press, 2003.

———. "SF: Science Fiction, Speculative Fabulation, String Figures, So Far." Pilgrim Award Acceptance Talk Presented at the Science Fiction Research Association (SFRA) Conference, Lublin, Poland, July 2011. people.ucsc.edu/~haraway/PilgrimAward.html (accessed January 9, 2016).

———. "'There Are Always More Things Going On Than You Thought!' Methodologies as Thinking Technologies: Interview with Donna Haraway." In *Bits of Life: Feminism at the Intersections of Media, Bioscience, and Technology*, edited by Anneke Smelik and Nina Lykke, 32–41. Seattle: University of Washington Press, 2008.

———. *When Species Meet*. Minneapolis: University of Minnesota Press, 2008.

Haraway, Donna, Noboru Ishikawa, Gilbert Scott, Kenneth Olwig, Anna. L. Tsing, and Nils Bubandt. "Anthropologists Are Talking—About the Anthropocene." *Ethnos* (2015): 1–30. dx.doi.org/10.1080/00141844.2015.1105838 (accessed January 9, 2016).

Harding, Stephan. *Animate Earth: Science, Intuition, and Gaia*. White River Junction: Chelsea Green Publishing Company, 2006.

Hardt, Michael, and Antonio Negri. *Commonwealth*. Cambridge: Harvard University Press, 2009.

———. *Empire*. Cambridge: Harvard University Press, 2001.

———. *Multitude: War and Democracy in the Age of Empire*. New York: Penguin, 2004.

Harman, Graham. *Guerilla Metaphysics: Phenomenology and the Carpentry of Things*. Chicago: Open Court, 2005.

———. "On the Undermining of Objects: Grant, Bruno, and Radical Philosophy." In *The Speculative Turn: Continental Materialism and Realism*, edited by Levi Bryant, Nick Srnicek, and Graham Harman, 21–40. Melbourne: re.press, 2011.

———. *Towards Speculative Realism: Essays and Lectures*. Winchester: Zero Books, 2010.

Harvey, Graham. *Animism: Respect for the Living World*. New York: Columbia University Press, 2005.

———. "An Animist Manifesto." *PAN: Philosophy, Activism, Nature* 9 (2012): 2–4. www.panjournal.net/issues/9 (accessed January 28, 2016).

———, ed. *The Handbook of Contemporary Animism*. New York: Routledge, 2014.

The Heart Sutra: The Womb of Buddhas. Translated and commentary by Red Pine. Berkeley: Counterpoint, 2004.

Heft, Harry. *Ecological Psychology in Context: James Gibson, Roger Barker, and the Legacy of William James' Radical Empiricism*. Mahwah: L. Erlbaum Associates, 2001.

Hegel, Georg Wilhelm Friedrich. *Hegel's Phenomenology of Spirit*. Translated by A. V. Miller. Oxford: Oxford University Press, 1977.

Heidegger, Martin. *Being and Time*. Translated by John Macquarrie and Edward Robinson. New York: Harper and Row, 1962.

———. *Contributions to Philosophy (From Enowning)*. Translated by Parvis Emad and Kenneth Maly. Bloomington: Indiana University Press, 1999.

———. *Country Path Conversations*. Translated by Brett Davis. Bloomington: Indiana University Press, 2010.

———. *The Fundamental Concepts of Metaphysics: World, Finitude, Solitude.* Translated by William McNeill and Nicholas Walker. Bloomington: Indiana University Press, 1995.

———. *An Introduction to Metaphysics.* Translated by Ralph Manheim. New Haven: Yale University Press, 1987.

———. "Letter on 'Humanism.'" In *Pathmarks,* edited by William McNeill, translated by Frank A. Capuzzi, 239–76. Cambridge: Cambridge University Press, 1998.

———. "The Origin of the Work of Art." In *Off the Beaten Track,* translated by Julian Young and Kenneth Haynes, 1–56. Cambridge: Cambridge University Press, 2002.

———. *Nietzsche: Volumes One and Two.* Translated by David Farrell Krell. San Francisco: Harper SanFrancisco, 1991.

———. *What Is Called Thinking?* Translated by J. Glenn Gray. New York: Perennial, 1976.

Heise, Ursula K. *Chronoschisms: Time, Narrative, and Postmodernism.* Cambridge: Cambridge University Press, 1997.

———. *Sense of Place, Sense of Planet: The Environmental Imagination of the Global.* Oxford: Oxford University Press, 2008.

Hiltner, Ken. *Milton and Ecology.* Cambridge: Cambridge University Press, 2003.

Holland, Nancy J., and Patricia Huntington, eds. *Feminist Interpretations of Martin Heidegger.* University Park: Pennsylvania State University Press, 2001.

Husserl, Edmund. *The Crisis of European Sciences and Transcendental Phenomenology: An Introduction to Phenomenological Philosophy.* Translated by David Carr. Evanston: Northwestern University Press, 1970.

Instone, Lesley. "Risking Attachment in the Anthropocene." In *Manifesto for Living in the Anthropocene,* edited by Katherine Gibson, Deborah Bird Rose, and Ruth Fincher. Brooklyn: Punctum Books, 2015.

Iovino, Serenella, and Serpil Oppermann, eds. *Material Ecocriticism.* Bloomington: Indiana University Press, 2014.

Irigaray, Luce. *Elemental Passions.* Translated by Joanne Collie and Judith Still. New York: Routledge, 1992.

———. *An Ethics of Sexual Difference.* Translated by Carolyn Burke and Gillian C. Gill. Ithaca: Cornell University Press, 1993.

———. *I Love to You: Sketch for a Felicity within History.* Translated by Alison Martin. New York: Routledge, 1996.

———. *Luce Irigaray: Key Writings.* Edited by Luce Irigaray. London: Continuum, 2004.

———. "The Question of the Other." Translated by Noah Guynn. *Yale French Studies* 87 (1995): 7–19.

———. *This Sex Which Is Not One.* Translated by Catherine Porter. Ithaca: Cornell University Press, 1985.

———. *Sharing the World.* Translated by Stephen Pluhacek. London: Continuum, 2008.

———. *The Way of Love.* Translated by Heidi Bostic and Stephen Pluhacek. London: Continuum, 2002.

———. "Women's Exile." Translated by Couze Venn. *Ideology and Consciousness* 1 (1977): 62–76.

Jaspers, Karl. *General Psychopathology, Volume 2.* Translated by J. Hoenig and Marian W. Hamilton. Baltimore: Johns Hopkins University Press, 1997.

———. *Man in the Modern Age.* Translated by Eden Paul and Cedar Paul. New York: Routledge, 2010.

———. *The Origin and Goal of History.* Translated by Michael Bullock. London: Routledge, 2010.

———. *Philosophy of Existence.* Translated by Richard J. Grabay. Philadelphia: Philadelphia University Press, 1971.

———. *Reason and Existenz: Five Lectures.* Translated by William Earle. Milwaukee: Marquette University Press, 1996.

———. *Socrates, Buddha, Confucius, Jesus: The Paradigmatic Individuals.* Translated by Ralph Manheim. San Diego: Harcourt Brace and Company, 1962.

———. *Von der Wahrheit*. Munich: R. Piper, 1947.

Jenkins, Willis, Mary Evelyn Tucker, and John Grim, eds. *Routledge Handbook of Religion and Ecology*. New York: Routledge, 2016.

Jonas, Hans. *The Imperative of Responsibility: In Search of an Ethics for the Technological Age*. Translated by Hans Jonas in collaboration with David Herr. Chicago: University of Chicago Press, 1984.

———. *The Phenomenon of Life: Toward a Philosophical Biology*. New York: Harper and Row, 1966.

Kafka, Franz. "The Cares of a Family Man." In *The Complete Stories*, edited by Nahum N. Glatzer, translated by Willa and Edwin Muir, 427–29. New York: Schocken, 1971.

———. "Couriers." In *Existentialism*, edited by Robert C. Solomon, 167. New York: Random House, 1974.

Kalupahana, David J. *Nāgārjuna: The Philosophy of the Middle Way*. Albany: SUNY Press, 1986.

Kauffman, Stuart. *At Home in the Universe: The Search for Laws of Self-Organization and Complexity*. Oxford: Oxford University Press, 1995.

Kaufmann, Walter, ed. *Existentialism: From Dostoevsky to Sartre*. Revised and expanded edition. New York: Penguin Books, 1988.

Keller, Catherine. *From a Broken Web: Separation, Sexism, and Self*. Boston: Beacon Press, 1986.

Kierkegaard, Søren. *Fear and Trembling*. Translated by Sylvia Walsh. Cambridge: Cambridge University Press, 2006.

———. *Papers and Journals: A Selection*. Translated by Alastair Hannay. London: Penguin Books, 1996.

———. *Philosophical Fragments*. Translated by Howard Hong and Edna Hong. Princeton: Princeton University Press, 1985.

———. *The Sickness unto Death*. Edited and translated by Howard V. Hong and Edna H. Hong. Princeton: Princeton University Press, 1980.

Klein, Naomi. *The Shock Doctrine: The Rise of Disaster Capitalism*. New York: Picador, 2007.

Kohn, Eduardo. *How Forests Think: Toward an Anthropology Beyond the Human*. Berkeley: University of California Press, 2013.

Kolsbun, Ken. *Peace: The Biography of a Symbol*. Washington, DC: National Geographic Books, 2008.

Latour, Bruno. "Angels without Wings: A Conversation Between Bruno Latour and Anselm Franke." In *Animism, Volume 1*, edited by Anselm Franke, 86–96. Berlin: Sternberg Press, 2011.

———. "The Anthropocene and the Destruction of the Image of the Globe." Fourth Lecture in the series *Facing Gaia: A New Enquiry into Natural Religion*. February 25, 2013. www.ed.ac.uk/schools-departments/humanities-soc-sci/news-events/lectures/gifford-lectures/archive/series-2012–2013/bruno-latour/lecture-four (accessed October 9, 2015).

———. "An Attempt at a 'Compositionist Manifesto.'" *New Literary History* 41 (2010): 471–90.

———. "A Cautious Prometheus? A Few Steps Toward a Philosophy of Design with Special Attention to Peter Sloterdijk." In *In Media Res: Peter Sloterdijk's Spherological Poetics of Being*, edited by Willem Schinkel and Liesbeth Noordegraaf-Eelens, 151–64. Amsterdam: Amsterdam Unversity Press, 2011.

———. *An Inquiry into Modes of Existence: An Anthropology of the Moderns*. Translated by Catherine Porter. Cambridge: Harvard University Press, 2013.

———. *Facing Gaia: Six Lectures on the Political Theology of Nature*. Gifford Lectures on Natural Religion. Edinburgh, February 18–28, 2013.

———. *Politics of Nature: How to Bring the Sciences into Democracy*. Translated by Catherine Porter. Cambridge: Harvard University Press, 2004.

———. "Why Has Critique Run Out of Steam? From Matters of Fact to Matters of Concern." *Critical Inquiry* 30.2 (2004): 225–48.

Le Guin, Ursula K. "The Carrier Bag Theory of Fiction." In *Dancing at the Edge of the World: Thoughts on Words, Women, Places*, 165–70. New York: Grove Press, 1989.
Léon, Céline, and Sylvia Walsh, eds. *Feminist Interpretations of Søren Kierkegaard*. University Park: Pennsylvania State University Press, 1997.
Leopold, Aldo. *A Sand County Almanac and Sketches Here and There*. London: Oxford University Press, 1968.
Levin, David Michael, ed. *Modernity and the Hegemony of Vision*. Berkeley: University of California Press, 1993.
Levinas, Emmanuel. *Entre Nous: Thinking-of-the-Other*. Translated by Michael B. Smith and Barbara Harshav. New York: Columbia University Press, 1998.
———. *Existence and Existents*. Translated by Alphonso Lingis. Dordrecht: Kluwer, 1988.
———. *Otherwise Than Being Or Beyond Essence*. Translated by Alphonso Lingis. Pittsburgh: Duquesne University Press, 2002.
———. "Revelation in the Jewish Tradition." In *The Levinas Reader*, edited by Sean Hand, 190–210. Oxford: Blackwell Publishers, 1997.
———. *Totality and Infinity: An Essay on Exteriority*. Translated by Alphonso Lingis. Dordrecht: Kluwer Academic Publishers, 1991.
Lingis, Alphonso. *Abuses*. Berkeley: University of California Press, 1994.
———. *The Community of Those Who Have Nothing in Common*. Bloomington: Indiana University Press, 1994.
———. *The Imperative*. Bloomington: Indiana University Press, 1998.
———. *Sensation: Intelligibility in Sensibility*. Atlantic Highlands: Humanities Press, 1996.
Llewelyn, John. *The Rigor of a Certain Inhumanity: Toward a Wider Suffrage*. Bloomington: Indiana University Press, 2012.
Lyotard, Jean-François. *The Inhuman: Reflections on Time*. Translated by Geoffrey Bennington and Rachel Bowlby. Stanford: Stanford University Press, 1991.
———. *The Postmodern Condition: A Report on Knowledge*. Translated by Geoff Bennington and Brian Massumi. Minneapolis: University of Minnesota Press, 1984.
Macy, Joanna. *World as Lover, World as Self: Courage for Global Justice and Ecological Renewal*. Berkeley: Parallax Press, 2007.
Malabou, Catherine. *Changing Difference: The Feminine and the Question of Philosophy*. Translated by Carolyn Shread. Cambridge: Polity Press, 2011.
———. *Plasticity at the Dusk of Writing: Dialectic, Destruction, Deconstruction*. Translated by Carolyn Shread. New York: Columbia University Press, 2010.
———. *What Should We Do with Our Brain?* Translated by Sebastian Rand. New York: Fordham University Press, 2008.
Malpas, Jeff. *Heidegger and the Thinking of Place: Explorations in the Topology of Being*. Cambridge: MIT Press, 2012.
Mandoki, Katya. *The Indispensable Excess of the Aesthetic: An Evolution of Sensibility in Nature*. Lanham: Lexington Books, 2015.
Marder, Michael. *Dust*. New York: Bloomsbury Academic, 2016.
———. *The Philosopher's Plant: An Intellectual Herbarium*. New York: Columbia University Press, 2014.
———. *Plant-Thinking: A Philosophy of Vegetal Life*. New York: Columbia University Press, 2013.
Marion, Jean-Luc. *In Excess: Studies of Saturated Phenomena*. Translated by Robyn Horner and Vincent Berraud. New York: Fordham University Press, 2002.
McKibben, Bill. *The End of Nature*. New York: Random House, 2006.
Merchant, Carolyn. *The Death of Nature: Women, Ecology, and the Scientific Revolution*. San Francisco: HarperSanFrancisco, 1990.
Merleau-Ponty, Maurice. *Nature: Course Notes from the Collège de France*. Translated by Robert Vallier. Evanston: Northwestern University Press, 2003.
———. *Phenomenology of Perception*. Translated by Colin Smith. New York: Routledge and Kegan Paul LTD, 1998.

———. *Signs*. Translated by Richard C. McCleary. Evanston: Northwestern University Press, 1964.
———. *The Visible and the Invisible*. Translated by Alphonso Lingis. Evanston: Northwestern University Press, 1968.
Metzger, Bruce M., and Roland E. Murphy, eds. *The New Oxford Annotated Bible with the Apocrypha*. New Revised Standard Version. New York: Oxford University Press, 1991.
Michaelsen, Scott, and David E. Johnson, eds. *At the Heart: Of Jean-Luc Nancy. A Special Issue of The New Centennial Review*, vol. 2, no. 3, Fall 2002. muse.jhu.edu/journals/new_centennial_review/toc/ncr2.3.html (accessed January 28, 2016).
Mickey, Sam. "Contributions to Anthropocosmic Environmental Ethics." *Worldviews: Environment, Culture, Religion* 11.2 (2007): 226–47.
———. *On the Verge of a Planetary Civilization: A Philosophy of Integral Ecology*. London: Rowman & Littlefield International, 2014.
———. "Review of *Becoming Animal: An Earthly Cosmology*, by David Abram." *Worldviews: Global Religions, Culture, and Ecology* 16.1 (2012): 99–101.
Miller, Adam. *Speculative Grace: Bruno Latour and Object-Oriented Theology*. New York: Fordham University Press, 2013.
Moe, Aaron M. *Zoopoetics: Animals and the Making of Poetry*. Lanham: Lexington Books, 2014.
Mortimer-Sandilands, Catriona, and Bruce Erickson, eds. *Queer Ecologies: Sex, Nature, Politics, Desire*. Indiana University Press, 2010.
Morton, Timothy. "The Best Eco Anthem," *Ecology without Nature* blog, October 29, 2014. ecologywithoutnature.blogspot.com/2014/10/the-best-eco-anthem.html (accessed January 28, 2016).
———. "Buddhaphobia: Nothingness and the Fear of Things." In *Nothing: Three Inquiries in Buddhism*, edited by Marcus Boon, Eric Cazdyn, and Timothy Morton. Chicago: University of Chicago Press, 2015.
———. *Dark Ecology: Toward a Logic of Future Coexistence*. New York: Columbia University Press, 2016.
———. "Deconstruction and/as Ecology." In *The Oxford Handbook of Ecocriticism*, edited by Greg Garrard, 291–304. Oxford: Oxford University Press, 2014.
———. *The Ecological Thought*. Cambridge: Harvard University Press, 2010.
———. *Ecology without Nature: Rethinking Environmental Aesthetics*. Cambridge: Harvard University Press, 2007.
———. "Elementality." In *Elemental Ecocriticism: Thinking with Earth, Air, Water, and Fire*, edited by Jeffrey Jerome Cohen and Lowell Duckert, 271–85. Minneapolis: University of Minnesota Press, 2015.
———. *Hyperobjects: Philosophy and Ecology after the End of the World*. Minneapolis: University of Minnesota Press, 2013.
———. "Guest Column: Queer Ecology." *PMLA* 125.2 (2010): 273–82.
———. *Realist Magic: Objects, Ontology, Causality*. Ann Arbor: Open Humanities Press, 2013.
———. "They Are Here." In *The Nonhuman Turn*, edited by Richard Grusin, 167–92. Minneapolis: University of Minnesota Press, 2015.
———. "Treating Objects Like Women: Feminist Ontology and the Question of Essence." In *International Perspectives in Feminist Ecocriticism*, edited by Greta Gaard, Simon C. Estok, and Serpil Oppermann, 56–69. London: Routledge, 2013.
———. "Wellek Lectures: The Description." *Ecology without Nature* blog, October 29, 2014. ecologywithoutnature.blogspot.com/2014/04/wellek-lectures-description.html?utm_source=twitterfeed&utm_medium=twitter (accessed January 28, 2016).
———. "X-Ray." In *Prismatic Ecology: Ecotheory Beyond Green*, edited by Jeffrey Jerome Cohen, 311–27. Minneapolis: University of Minnesota Press, 2013.
Myers, Ella. *Worldly Ethics: Democratic Politics and Care for the World*. Durham: Duke University Press, 2013.

Nagarjuna. *The Fundamental Wisdom of the Middle Way*. Translated and commentary by Jay Garfield. Oxford: Oxford University Press, 1995.

Nancy, Jean-Luc. *Adoration: The Deconstruction of Christianity II*. Translated by John McKeane. New York: Fordham University Press, 2013.

———. *Being Singular Plural*. Translated by Robert D. Richardson and Anne E. O'Byrne. Stanford: Stanford University Press, 2000.

———. *Corpus*. Translated by Richard A. Rand. New York: Fordham University Press, 2008.

———. *The Creation of the World* or *Globalization*. Translated by François Raffoul and David Pettigrew. Albany: SUNY Press, 2007.

———. *Dis-Enclosure: The Deconstruction of Christianity*. Translated by Bettina Bergo, Gabriel Malenfant, and Michael B. Smith. New York: Fordham University Press, 2008.

———. *The Experience of Freedom*. Translated by Bridget McDonald. Stanford: Stanford University Press, 1993.

———. *The Inoperative Community*. Edited by Peter Connor. Translated by Peter Connor, Lisa Garbus, Michael Holland, and Simona Sawhney. Minneapolis: University of Minnesota Press, 1991.

———. *Intoxication*. Translated by Philip Armstrong. New York: Fordham University Press, 2016.

———. *Listening*. Translated by Charlotte Mandell. New York: Fordham University Press, 2007.

———. *Multiple Arts: The Muses II*. Edited by Simon Sparks. Stanford: Stanford University Press, 2006.

———. *The Muses*. Translated by Peggy Kamuf. Stanford: Stanford University Press, 1996.

———. *Noli Me Tangere: On the Raising of the Body*. Translated by Sarah Clift, Pascale-Anne Brault, and Michael Naas. New York: Fordham University Press, 2008.

———. "The Real Outside Is 'At the Heart' of the Inside: Interview." *ATOPIA* (2007). www.rave.ca/en/journals_info/1432/moohk (accessed January 28, 2016).

———. *The Sense of the World*. Translated by Jeffrey Librett. Minneapolis: University of Minnesota Press, 1997.

Nancy, Jean-Luc, and Aurélien Barrau. *What's These Worlds Coming To?* Translated by Travis Holloway and Flor Méchain. New York: Fordham University Press, 2015.

Nhat Hanh, Thich. *Interbeing: Fourteen Guidelines for Engaged Buddhism*. Berkeley: Parallax Press, 1998.

Niemeier, Susanne. "To Be in Control: Kind-Hearted and Cool-Headed: The Head-Heart Dichotomy in English." In *Culture, Body, and Language: Conceptualizations of Internal Body Organs Across Cultures and Languages*, edited by Farzad Sharifian, René Dirven, Ning Yu, and Susanne Niemeier, 349–72. Berlin: Mouton de Gruyter, 2008.

Nietzsche, Friedrich. *The Birth of Tragedy*. Translated by Ronald Speirs. Cambridge: Cambridge University Press, 1999.

———. *Daybreak: Thoughts on the Prejudices of Morality*. Translated by R. J. Hollingdale. Cambridge: Cambridge University Press, 2000.

———. *Ecce Homo: How to Become What One Is*. Translated by R. J. Hollingdale. New York: Penguin Books, 1992.

———. *The Gay Science*. Translated by Walter Kaufmann. New York: Vintage Books, 1974.

———. *Thus Spoke Zarathustra: A Book for Everyone and No One*. Translated by R. J. Hollingdale. New York: Penguin Books, 1969.

———. *Twilight of the Idols*. Translated by Richard Polt. Indianapolis: Hackett Publishing, 1997.

———. *The Will to Power*. Translated by Walter Kaufmann and R. J. Hollingdale. New York: Vintage Books, 1968.

———. *Writing from the Late Notebooks*. Edited by Rüdiger Bittner. Cambridge: Cambridge University Press, 2003.

Nishitani, Keiji. *Religion and Nothingness*. Translated by Jan van Bragt. Berkeley: University of California Press, 1983.

———. *The Self-Overcoming of Nihilism*. Translated by Graham Parkes with Setsuko Aihara. Albany: SUNY Press, 1990.

Oliver, Kelly. *Earth and World: Philosophy after the Apollo Missions*. New York: Columbia University Press, 2015.

Oliver, Kelly, and Marilyn Pearsall, eds. *Feminist Interpretations of Friedrich Nietzsche*. University Park: Pennsylvania State University Press, 1998.

Otto, Rudolf. *The Idea of the Holy*. Translated by John W. Harvey. London: Oxford University Press, 1958.

Phillips, Mary, and Nick Rumens, eds. *Contemporary Perspectives on Ecofeminism*. London: Routledge, 2015.

Plato. *Apology*. Translated by Harold North Fowler. Loeb Classical Library 36. Cambridge: Harvard University Press, 2001.

———. *Phaedrus*. Translated by Albert A. Anderson. Millis: Agora Publications, 2009.

———. *Timaeus*. Translated by R. G. Bury. Cambridge: Harvard University Press, 2005.

———. *The Republic*. Translated by Allan Bloom. New York: Basic Books, 1968.

Plumwood, Val. *Feminism and the Mastery of Nature*. London: Routledge, 1993.

Polt, Richard. *The Emergency of Being: On Heidegger's* Contributions to Philosophy. Ithaca: Cornell University Press, 2006.

———. "The Event of Enthinking the Event." In *Companion to Heidegger's* Contributions to Philosophy, edited by Charles E. Scott, Susan M. Schoenbohm, Daniela Vallega-Neu, and Alejandro Vallega, 81–104. Bloomington: Indiana University Press, 2001.

Purdy, Jedidiah. *After Nature: A Politics for the Anthropocene*. Cambridge: Harvard University Press, 2015.

Riedlinger, Thomas. "Two Classic Trips: Jean-Paul Sartre and Adelle Davis." In *Hallucinogens: A Reader*, edited by Charles Grob, 47–63. New York: Jeremy P. Tarcher/Putnam, 2002.

Rilke, Rainer Maria. "Archaic Torso of Apollo." Translated by Edward Snow. In *The Art of the Sonnet*, edited by Stephen Burt and David Mikics, 230. Cambridge: Harvard University Press, 2010.

———. "A Walk." In *The Winged Energy of Delight: Selected Translations*, edited and translated by Robert Bly, 176. New York: Perennial, 2005.

Rinella, Michael. *Pharmakon: Plato, Drug Culture, and Identity in Ancient Athens*. Lanham: Lexington Books, 2011.

Rockström, Johan, et al. "Planetary Boundaries: Exploring the Safe Operating Space for Humanity." *Ecology and Society* 14.2 (2009). www.ecologyandsociety.org/vol14/iss2/art32 (accessed January 28, 2016).

Rockström, Johan, and Mattias Klum. *Big World, Small Planet: Abundance within Planetary Boundaries*. New Haven: Yale University Press, 2015.

Rose, Deborah Bird. *Dingo Makes Us Human: Life and Land in Australian Aboriginal Culture*. Cambridge: Cambridge University Press, 2000.

———. "The Ecological Humanities." In *Manifesto for Living in the Anthropocene*, edited by Katherine Gibson, Deborah Bird Rose, and Ruth Fincher, 1–5. Brooklyn: Punctum Books, 2015.

———. *Nourishing Terrains: Australian Aboriginal Views of Landscape and Wilderness*. Canberra: Australian Heritage Commission, 1996.

———. *Reports from a Wild Country: Ethics for Decolonisation*. Sydney: University of New South Wales, 2004.

———. "Val Plumwood's Philosophical Animism: Attentive Interactions in the Sentient World." *Environmental Humanities* 3 (2013): 93–109.

———. *Wild Dog Dreaming: Love and Extinction*. Charlottesville: University of Virginia Press, 2011.

Rousseau, Jean-Jacques. *On the Origin of Language: Two Essays, Jean-Jacques Rousseau and Johann Gottfried Herder*. Translated by John H. Moran and Alexander Gode. Chicago: University of Chicago Press, 1986.

Rubenstein, Mary-Jane. *Worlds without End: The Many Lives of the Multiverse*. New York: Columbia University Press, 2014.

Ruether, Rosemary Radford. *Women and Redemption: A Theological History*. 2nd edition. Minneapolis: Fortress Press, 2012.

Sartre, Jean-Paul. *Being and Nothingness: A Phenomenological Essay on Ontology*. Translated by Hazel E. Barnes. New York: Washington Square Press, 1984.

———. *Existentialism Is a Humanism*. Translated by Carol Macomber. New Haven: Yale University Press, 2007.

———. *Nausea*. Translated by Lloyd Alexander. New York: New Directions, 1964.

Scranton, Roy. *Learning to Die in the Anthropocene: Reflections on the End of a Civilization*. San Francisco: City Lights Books, 2015.

Seidell, Streeter. *White Whine: A Study of First-World Problems*. Avon: Adams Media, 2013.

Shaviro, Steven. *The Universe of Things: On Speculative Realism*. Minneapolis: University of Minnesota Press, 2014.

Shea, Louisa. *The Cynic Enlightenment: Diogenes in the Salon*. Baltimore: Johns Hopkins University Press, 2010.

Simmer-Brown, Judith. *Dakini's Warm Breath: The Feminine Principle in Tibetan Buddhism*. Boston: Shambhala Publications, 2001.

Slewers, Alfred Kentigern, ed. *Re-Imagining Nature: Environmental Humanities and Ecosemiotics*. Lanham: Bucknell University Press, 2014.

Sloterdijk, Peter. *Bubbles: Spheres Volume I: Microspherology*. Translated by Wieland Hoban. Los Angeles: Semiotext(e), 2011.

———. *Critique of Cynical Reason*. Translated by Michael Eldred. Minneapolis: University of Minnesota Press, 1987.

———. *Globes: Spheres Volume II: Macrospherology*. Translated by Wieland Hoban. Los Angeles: Semiotext(e), 2014.

———. *In the World Interior of Capital: For a Philosophical Theory of Globalization*. Translated by Wieland Hoban. Malden: Polity Press, 2013.

———. *Nietzsche Apostle*. Translated by Steven Corcoran. Los Angeles: Semiotext(e), 2013.

———. *Philosophical Temperaments: From Plato to Foucault*. Translated by Thomas Dunlap. New York: Columbia University Press, 2013.

———. *Sphären III: Schäume, Plurale Sphärologie*. Berlin: Suhrkamp, 2004.

———. *You Must Change Your Life: On Anthropotechnics*. Translated by Wieland Hoban. Malden: Polity Press, 2013.

Slovic, Scott, Swarnalatha Rangarajan, and Vidya Sarveswaran, eds. *Ecoambiguity, Community, and Development: Toward a Politicized Ecocriticism*. Lanham: Lexington Books, 2014.

Snider, John. "I Heart Huckabees." *Philosophy Now: A Magazine of Ideas* 51, June/July 2005. philosophynow.org/issues/51/I_Heart_Huckabees (accessed September 3, 2015).

Snyder, Gary. *The Practice of the Wild*. Washington, DC: Shoemaker & Hoard, 1990.

Sparrow, Tom. *Plastic Bodies: Rebuilding Sensation after Phenomenology*. London: Open Humanities Press, 2015.

Spitzner, Manfred, et al. "Increased Activation of Indirect Semantic Associations under Psilocybin." *Biological Psychiatry* 39.12 (1996): 1055–57.

Steinbock, Anthony J. *Phenomenology and Mysticism: The Verticality of Religious Experience*. Bloomington: Indiana University Press, 2007.

Taylor, Mark C. *After God*. Chicago: University of Chicago Press, 2009.

Thacker, Eugene. *In the Dust of This Planet: Horror of Philosophy, Vol. 1*. Winchester: Zero Books, 2011.

———. *Starry Speculative Corpse: Horror of Philosophy, Vol. 2*. Winchester: Zero Books, 2015.

———. *Tentacles Longer Than Night: Horror of Philosophy, Vol. 3*. Winchester: Zero Books, 2015.

Thompson, Hunter S. "A Man Has to Be Something, He Has to Matter: Hunter S. Thompson to Hume Logan." In *Letters of Note: An Eclectic Collection of Correspondence Deserving of a Wider Audience*, edited by Shaun Usher, 64–67. San Francisco: Chronicle Books, 2014.

Tillich, Paul. *The Courage To Be*. 2nd edition. New Haven: Yale University Press, 2000.

———. *Systematic Theology: Volume 1*. Chicago: University of Chicago Press, 1951.

Toadvine, Ted. "Limits of the Flesh: The Role of Reflection in David Abram's Ecophenomenology." *Environmental Ethics* 27.2 (2005): 155–70.

Trigg, Dylan. *The Thing: A Phenomenology of Horror*. Winchester: Zero Books, 2014.

Trismegistus, Hermes. "*The Emerald Tablet* (*Tabula Smaragdina*)." In *The Alchemy Reader: From Hermes Trismegistus to Isaac Newton*, edited by Stanton J. Linden, 27–28. Cambridge: Cambridge University Press, 2003.

Tu, Weiming. "The Ecological Turn in New Confucian Humanism: Implications for China and the World." *Daedalus* 130.4 (2001): 243–64.

Viveiros de Castro, Eduardo. *The Relative Native: Essays on Indigenous Conceptual Worlds*. Chicago: Hau Books, 2015.

Waldenfels, Bernhard. *Phenomenology of the Alien: Basic Concepts*. Translated by Tanja Stahler and Alexander Kozin. Evanston: Northwestern University Press, 2011.

Wall, Thomas Carl. *Radical Passivity: Levinas, Blanchot, and Agamben*. Albany: SUNY Press, 1999.

Walsh, Roger, and Charles S. Grob. "Conclusion: What Did These Elders Learn, and What Can We Learn From Them?" In *Higher Wisdom: Eminent Elders Explore the Continuing Impact of Pyshcedelics*, edited by Roger Walsh and Charles S. Grob, 241–58. Albany: SUNY Press, 2005.

Wasson, Robert Gordon, Albert Hofmann, and Carl A. P. Ruck. *The Road to Eleusis: Unveiling the Secret of the Mysteries*. 30th anniversary edition. Berkeley: North Atlantic Books, 2008.

Watts, Alan. *Eastern Wisdom, Modern Life: Collected Talks, 1960–1969*. Novato: New World Library, 2011.

Webster, David. *Dispirited: How Spirituality Makes Us Stupid, Selfish, and Unhappy*. Winchester: Zero Books, 2012.

Whitehead, Alfred North. *Process and Reality: An Essay in Cosmology*. Corrected edition. Edited by David Ray Griffin and Donald W. Sherburne. New York: Free Press, 1978.

Wood, David. *Time after Time*. Bloomington: Indiana University Press, 2007.

Wood, Robert E. "The Heart in Heidegger's Thought." *Continental Philosophy Review* 48.4 (2015): 445–62.

Zimmerman, Michael E. *Contesting Earth's Future: Radical Ecology and Postmodernity*. Berkeley and Los Angeles: University of California Press, 1994.

Žižek, Slavoj. *On Belief*. London: Routledge, 2001.

———. *In Defense of Lost Causes*. London: Verso, 2008.

Index

About the Author

Sam Mickey, PhD, is an adjunct professor at the University of San Francisco and the California Institute of Integral Studies, teaching and writing at the intersection of philosophy, religion, and ecology. Sam has also taught at Pacifica Graduate Institute and Dominican University of California, and he worked for several years at the Forum on Religion and Ecology at Yale. He is the author of *On the Verge of a Planetary Civilization: A Philosophy of Integral Ecology* (2014), and *Whole Earth Thinking and Planetary Coexistence: Ecological Wisdom at the Intersection of Religion, Ecology, and Philosophy* (2016). He is a coeditor with Sean Kelly and Adam Robbert of *The Variety of Integral Ecologies: Nature, Culture, and Knowledge in the Planetary Era* (upcoming). He blogs at BecomingIntegral.com.